Selected Titles in This Series

昭和57年 3 月31日文部省検定済
高等学校数学科用

基礎解析

小平邦彦 編

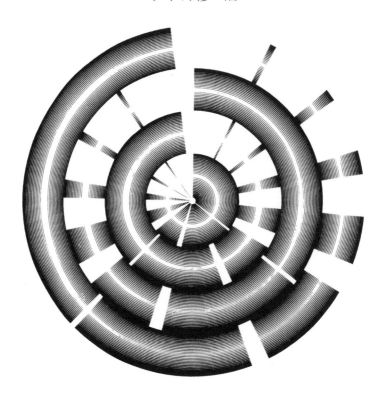

東京書籍株式会社

Mathematical World • Volume 11

Basic Analysis
Japanese Grade 11

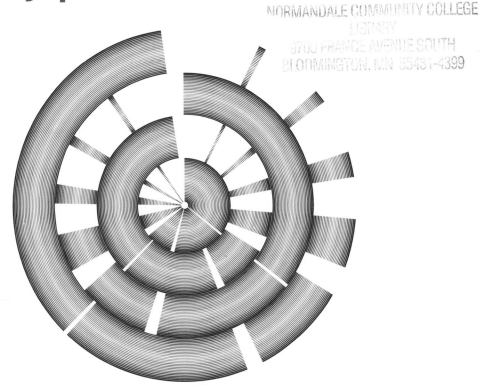

Kunihiko Kodaira, Editor

Hiromi Nagata, Translator
George Fowler, Translation Editor

American Mathematical Society
The University of Chicago School Mathematics Project

The University of Chicago School Mathematics Project

Zalman Usiskin, Director

Izaak Wirszup, Director, Resource Development Component

The translation and publication of this book
were made possible by the generous support of
The Amoco Foundation, Inc.

Translated by Hiromi Nagata
Translation edited by George Fowler

1991 *Mathematics Subject Classification.* Primary 00-01.

Library of Congress Cataloging-in-Publication Data
Kiso kaiseki. English.
 Basic analysis : Japanese grade 11 / edited by Kunihiko Kodaira; Hiromi Nagata, translator;
George Fowler, translation editor.
  p. cm. — (Mathematical world, ISSN 1055-9426; v. 11)
 Includes index.
 ISBN 0-8218-0580-0 (alk. paper)
 1. Mathematical analysis. I. Kodaira, Kunihiko, 1915– . II. Title. III. Series.
QA300.K52513 1996
515—dc20

96-23130
CIP

Textbook Series Preface

The University of Chicago School Mathematics Project

This textbook is part of a series of foreign mathematics texts that have been translated by the Resource Development Component of the University of Chicago School Mathematics Project (UCSMP). Established in 1983 with major funding from the Amoco Foundation, UCSMP has been since that time the nation's largest curriculum development and implementation project in school mathematics. The international focus of its resource component, together with the project's publication experience, makes UCSMP well suited to disseminate these remarkable materials.

The textbooks were originally translated to give U.S. educators and researchers a first-hand look at the content of mathematics instruction in educationally advanced countries. More specifically, they provided input for UCSMP as it developed new instructional strategies, textbooks, and materials of its own; the resource component's translations of over 40 outstanding foreign school mathematics publications, including texts, workbooks, and teacher aids, have been used in UCSMP–related research and experimentation and in the creation of innovative textbooks.

The resource component's translations include the entire mathematics curriculum (grades 1–10) used in the former Soviet Union, standard Japanese texts for grades 7–11, and innovative elementary textbooks from Hungary and Bulgaria.

The content of Japan's compulsory national curriculum for grades 7–11 is made available for the first time in English, thanks in part to the generosity of the Japanese publisher, Tokyo Shoseki Company, Ltd., which provided the copyright permissions.

Japanese Secondary School Mathematics Textbooks

The achievement of Japanese elementary and secondary students gained world prominence largely as a result of their superb performance in the International Mathematics Studies conducted by the International Association for the Evaluation of Educational Achievement. The Second International Mathematics Study surveyed mathematics achievement in 24 countries in 1981–82 and released its findings in 1984. The results are recapitulated in a 1987 national report entitled "The Underachieving Curriculum: Assessing U.S. School Mathematics from an International Perspective" (A National Report on the Second International Mathematics Study, 1987).

Let us take a brief look at the schooling behind much of Japan's economic success. The Japanese school system consists of a six-year primary school, a three-year lower secondary school, and a three-year upper secondary school. The first nine grades are compulsory, and enrollment is now 99.99%. According to 1990 statistics, 95.1% of age-group children are enrolled in upper secondary school, and the dropout rate is 2.2%. In terms of achievement, a typical Japanese student graduates from secondary school with roughly four more years of education than an average American high school graduate. The level of mathematics training achieved by Japanese students can be inferred from the following data:

Japanese Grade 7 Mathematics (New Mathematics 1) explores integers, positive and negative numbers, letters and expressions, equations, functions and proportions, plane figures, and figures in space. Chapter headings in *Japanese Grade 8 Mathematics* include calculating expressions, inequalities, systems of equations, linear functions, parallel lines and congruent figures, parallelograms, similar figures, and organizing data. *Japanese Grade 9 Mathematics* covers square roots, polynomials, quadratic equations, functions, circles, figures and measurement, and probability and statistics. The material in these three grades (lower secondary school) is compulsory for all students.

The textbook *Japanese Grade 10 Mathematics 1* covers material that is compulsory. This course, which is completed by over 97% of all Japanese students, is taught four hours per week and comprises algebra (including quadratic functions, equations, and inequalities), trigonometric functions, and coordinate geometry.

Japanese Grade 11 General Mathematics 2 is intended for the easier of the electives offered in that grade and is taken by about 40% of the students. It covers probability and statistics; vectors; exponential, logarithmic, and trigonometric functions; and differentiation and integration in an informal presentation.

The other 60% of students in grade 11 concurrently take two more extensive courses using the texts *Japanese Grade 11 Algebra and Geometry* and *Japanese Grade 11 Basic Analysis*. The first consists of fuller treatments of plane and solid coordinate geometry, vectors, and matrices. The second includes a more thorough treatment of trigonometry and an informal but quite extensive introduction to differential and integral calculus.

Some 25% of Japanese students continue with mathematics in grade 12. These students take an advanced course using the text *Probability and Statistics* and a second rigorous course with the text *Differential and Integral Calculus*.

One of the authors of these textbooks is Professor Hiroshi Fujita, who spoke at UCSMP's International Conferences on Mathematics Education in 1985, 1988, and 1991. Professor Fujita's paper on Japanese mathematics education appeared in *Developments in School Mathematics Education Around the World*, volume 1 (NCTM, 1987). The current school mathematics reform in Japan is described in the article "The Reform of Mathematics Education at the Upper Secondary School (USS) Level in Japan" by Professors Fujita, Tatsuro Miwa, and Jerry Becker in the proceedings of the Second International Conference, volume 2 of *Developments*.

Acknowledgments

It goes without saying that a publication project of this scope requires the commitment and cooperation of a broad network of institutions and individuals. In acknowledging their contributions, we would like first of all to express our deep appreciation to the Amoco Foundation. Without the Amoco Foundation's generous long-term support of the University of Chicago School Mathematics Project these books might never have been translated for use by the mathematics education community.

We are grateful to UCSMP Director Zalman Usiskin for his help and advice in making these valuable resources available to a wide audience at low cost. Robert Streit, Manager of the Resource Development Component, did an outstanding job coordinating the translation work and collaborating on the editing of most of the manuscripts. George Fowler, Steven R. Young, and Carolyn J. Ayers made a meticulous review of the translations, while Susan Chang and her technical staff at UCSMP handled the text entry and layout with great care and skill. We gratefully acknowledge the dedicated efforts of the translators and editors whose names appear on the title pages of these textbooks.

Izaak Wirszup, Director
UCSMP Resource Development Component

FOREWORD TO THE JAPANESE EDITION

This textbook is intended for students who study basic analysis after completing the study of Mathematics 1 in grade 10.

Mathematics was originally linked with science and technology; however, it gradually became independent of science and technology, and present-day mathematicians think freely about virtually everything possible. Therefore, mathematics is said to be a free creation of the human spirit.

On the other hand, mathematics studies mathematical principles which lie behind phenomena in various other fields, as we mentioned in the Foreword to the Mathematics 1 textbook. Therefore, mathematics is useful because it can be applied to other fields. Mathematics is a very important discipline; it is basic for the study of various other sciences.

In Chapter I, Section 1, you will learn about roots and exponential functions and their properties. Logarithmic functions are the inverse of exponential functions. In Section 2, you will study logarithmic functions and their properties.

In Chapter II we will investigate trigonometric functions. Trigonometric functions are an extension of trigonometric ratio, which you studied in Mathematics 1. You will learn about general angles, trigonometric functions and their properties, and the graphs of trigonometric functions in Section 1. The addition theorem for trigonometric functions, which you will encounter in Section 2, is very important.

A progression is a sequence of numbers arranged in accordance with a certain rule. In Chapter III, you will study arithmetic and geometric progressions. Mathematical induction, which you will learn about in Section 2, is the most basic method of mathematical prrof.

Differentiation enables us to find the function which represents the rate of change of a function, while conversely, integration enables us to find the original function, given the funciton representing the rate of change. As a result of the application of differentiation and integration, originated by Newton and Leibniz, science and technology made remarkable progress after the eighteenth century. You will study limits of functions, and basic differentiation and simple applications in Chapter IV.

In Chapter V, you will study integration and its application. For example, in Section 2 you will investigate the calculation of area and volume as an application of integration.

You cannot master mathematics by merely reading books and memorizing; you should think through the material, do calculations, draw figures, and solve problems by yourself. You cannot master swimming by reading books about swimming; you must swim in the water. Similarly, in order to master mathematics, you must think about mathematics by yourself.

TABLE OF CONTENTS

To the Student

Example
This marker designates a concrete example to help you understand the text.

Demonstration
This heading precedes a standard problem for better understanding of the material. Boxes labeled **[Solution]** and **[Proof]** give model answers.

Note:
This marker indicates an explanation to help you understand a particular point.

Problem 1
Problems for rapid mastery of current material and for introducing new material appear in the text with this label.

Exercises
At the end of each section problems are provided for practice with the material in that section.

Chapter Exercises
At the end of each chapter problems are provided for review of the entire chapter and practical application of the material. A problems involve primarily basic material, while $\mathbb{B}$ problems are a bit more advanced.

Reference
Refers to topics from other elective courses.

CHAPTER 1

EXPONENTIAL AND LOGARITHMIC FUNCTIONS

SECTION 1.   EXPONENTIAL FUNCTIONS
SECTION 2.   LOGARITHMIC FUNCTIONS

At one time logarithmic calculation was enthusiastically embraced because it reduced the multiplication and division of multi-digit numbers to addition and subtraction; it so reduced the burden of calculations that scientists and engineers said it lengthened their life span by ten years. However, logarithmic calculation is already a relic of the past, because the spread of various kinds of computers, including desktop calculators, has enabled us to obtain the results of calculations instantly.

But logarithmic functions and exponential functions, on which logarithmic functions are based, have become more important. In an exponential function, when the variable increases at a constant arithmetic rate, the value of the function increases at a geometric rate. Thus, an exponential function can describe fundamental phenomena developing over time. At present, exponential functions play an indispensable role in social sciences such as economics and psychology.

EXPONENTIAL FUNCTIONS

 ## The Function $y = x^n$

You have already learned about the functions $y = x$ and $y = x^2$. In this section, let's examine the properties and especially the graph of the function

$$y = f(x) = x^n \tag{1}$$

when n is a positive integer. First, let us consider it on the interval

$$x \geq 0. \tag{2}$$

Since

$$f(0) = 0^n = 0, \quad f(1) = 1^n = 1,$$

the graph of (1) passes through the origin and the point (1, 1). If we draw the graph of $y = f(x)$ on the interval in (2) for $n = 3$ and $n = 4$ by finding the value of $f(x)$ for the other values of x, we obtain the following graphs.

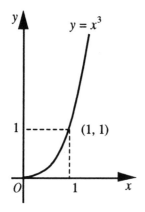
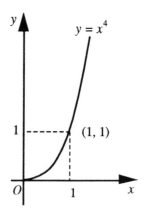

Problem 1 Graph $y = x^3$ $(x \geq 0)$ by finding the value of y for several values of x.

As in the graphs above, the value of $f(x)$ increases along with x on the interval $x > 0$. Let's check this fact by means of calculations.

First, since any positive number greater than 1 remains greater than 1 regardless of how many times it is multiplied by itself, and since if $0 < a < b$, then $\dfrac{b}{a} > 1$, we have

$$(\frac{b}{a})^n = \frac{b^n}{a^n} > 1.$$

Thus,

$$a^n < b^n.$$

So, for positive numbers a and b,

$$\text{if } a < b, \text{ then } f(a) < f(b).$$

Therefore, for $x > 0$, $f(x)$ increases with respect to x. So, for $x > 0$, $f(x)$ is an **increasing function**.

Moreover, the value of $f(x)$ grows without limit as x increases. Therefore, the range of $f(x)$ for $x \geq 0$ is $y \geq 0$.

Next, let's consider the graph of $y = f(x) = x^n$ if the domain of x is the set of all real numbers.

For an even number n, $(-a)^n = a^n$ for any arbitrary real number a. Therefore,

$$f(-a) = f(a). \tag{3}$$

So the graph is symmetric with respect to the y-axis.

For an odd number n, $(-a)^n = -a^n$ for any arbitrary real number a. Therefore,

$$f(-a) = -f(a). \tag{4}$$

So the graph is symmetric with respect to the origin.

We can use these facts to graph $y = f(x) = x^n$ when the domain is the set of all real numbers based on the graph of $x \geq 0$.

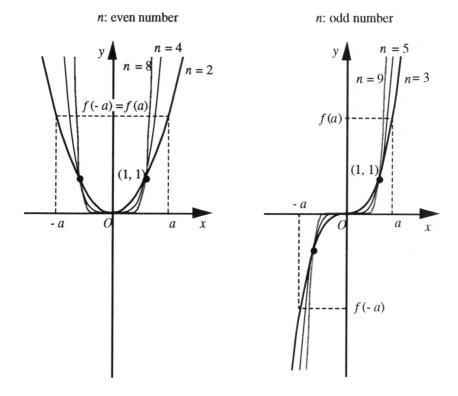

In general, a function to which property (3) applies is called an **even function**, and a function to which property (4) applies is an **odd function**.

The function $y = x^n$ is an even function for an even number n and an odd function for an odd number n.

(Problem 2)   State the range of $y = x^n$, taking the domain of x as the set of all real numbers.

2 nth Roots

Given a real number a and a positive integer n, the number x satisfying

$$x^n = a$$

is called the **nth root** of a .

A square root is a second root and a cubic root is a third root. A square root, a cubic root, ..., an nth root can all simply be called **roots**. In this section, we will consider the nth root of real numbers.

The nth root of a is the value of x for $y = a$ in the function $y = x^n$. Let's examine this function by means of a graph.

(1) When n is an even number.

   (i)   For $a > 0$, a has nth roots. We will write the positive root as $\sqrt[n]{a}$. The negative root can be expressed as $-\sqrt[n]{a}$ in view of the symmetry of the graph.

   (ii) For $a = 0$, 0 is the only nth root of a. We will also write it as $\sqrt[n]{a}$.

   (iii) For $a < 0$, a has no nth roots in the realm of real numbers.

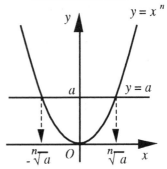

n: an even number

(Note:) As you have already learned, $\sqrt[n]{a}$ is equivalent to $\sqrt[2]{a}$.

(Example 1) $\sqrt[4]{81} = 3,\ \sqrt[6]{64} = 2$

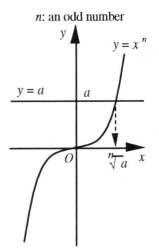

n: an odd number

(2) When a is an odd number.

In this case, a has only a single nth root, and the sign of a makes no difference; the sign of the root is the same as the sign of a. We will write this root as $\sqrt[n]{a}$.

Example 2 $\sqrt[3]{8} = 2,\ \sqrt[5]{-32} = -2$

Example 3   For either an odd number n or an even number n,

$$\sqrt[n]{0} = 0,\ \sqrt[n]{1} = 1.$$

Problem 1   Find all the following real number roots:

        (1) 4th roots of 16   (2) 4th roots of -16

        (3) 3rd roots of 27   (4) 3rd roots of -27

The following formulas hold for nth roots.

The Formulas of nth Roots

For $a > 0$ and $b > 0$,

$$[\mathrm{I}]\quad \sqrt[n]{a}\,\sqrt[n]{b} = \sqrt[n]{ab}\ ;\qquad\qquad [\mathrm{II}]\quad \frac{\sqrt[n]{a}}{\sqrt[n]{b}} = \sqrt[n]{\frac{a}{b}}.$$

Let's prove formula [I].

Multiplying the left side by itself n times, we obtain

$$\left(\sqrt[n]{a}\,\sqrt[n]{b}\right)^n = \left(\sqrt[n]{a}\right)^n\left(\sqrt[n]{b}\right)^n = ab.$$

Therefore, the nth power of $\sqrt[n]{a}\,\sqrt[n]{b}$ is equal to ab.

    On the other hand, a positive number ab has only a single positive nth root, $\sqrt[n]{ab}$. Thus,

$$\sqrt[n]{a}\,\sqrt[n]{b} = \sqrt[n]{ab}.$$

Formula [II] can be proved analogously.

Problem 2 Simplify the following expressions:

(1) $\sqrt[4]{3}\,\sqrt[4]{27}$

(2) $\sqrt[3]{200} \div \sqrt[3]{25}$

(3) $\sqrt[3]{0.01}\,\sqrt[3]{0.1}$

Example 4 $\sqrt[3]{40} = \sqrt[3]{2^3 \times 5} = \sqrt[3]{2^3}\,\sqrt[3]{5} = 2\sqrt[3]{5}$

Problem 3 Rewrite the following expressions as in Example 4.

(1) $\sqrt[3]{54}$

(2) $\sqrt[4]{48}$

(3) $\sqrt[3]{500}$

The following formula also holds for nth roots.

For $a > 0$, and positive integers m and n,

[III] $(\sqrt[n]{a})^m = \sqrt[n]{a^m}$.

Problem 4 Prove [III] by showing that the nth power of $(\sqrt[n]{a})^m$ is equal to a^m.

Problem 5 Simplify the following expressions:

(1) $(\sqrt[4]{25})^6$

(2) $\sqrt[3]{27^4}$

If n is a positive integer, the inverse function of $y = x^n$ $(x \geq 0)$ is $y = \sqrt[n]{x}$.

Therefore, the graph of $y = \sqrt[n]{x}$ can be created by reflecting the graph of $y = x^n$ $(x \geq 0)$ with respect to the straight line $y = x$. It is illustrated by the solid line in the figure to the right. Hence, we can see that

$y = \sqrt[n]{x}$ is an increasing function of x .

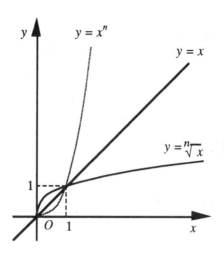

Comparison of nth Roots

For a positive integer n,

$$\text{if } 0 < a < b, \text{ then } \sqrt[n]{a} < \sqrt[n]{b} .$$

 Extension of Exponents

You have already learned in Mathematics I that the following laws of exponents hold when m and n are integers.

(1) $\quad a^m \times a^n = a^{m+n}$ (2) $\quad a^m \div a^n = a^{m-n}$

(3) $\quad (a^m)^n = a^{mn}$ (4) $\quad (ab)^m = a^m b^m$

(5) $\quad \left(\dfrac{a}{b}\right)^m = \dfrac{a^m}{b^m}$

As special cases, $a^0 = 1$ and $a^{-m} = \dfrac{1}{a^m}$.

Now let's define the meaning of a^r for $a > 0$ when r is any rational number.

If we assume that law (3) of exponents also holds for exponents of rational numbers, and for an arbitrary integer m and a positive integer n, then we have

$$\left(a^{\frac{m}{n}}\right)^n = a^{\frac{m}{n} \times n} = a^m.$$

Thus, $a^{\frac{m}{n}}$ is an nth root of a^m.

If we assume $a^{\frac{m}{n}} > 0$, then $a^{\frac{m}{n}} = \sqrt[n]{a^m}$. Therefore, we can formulate the following definition.

The Meaning of $a^{\frac{m}{n}}$

For $a > 0$, an arbitrary integer m, and a positive integer n,

$$a^{\frac{m}{n}} = \sqrt[n]{a^m}.$$

As a special case,

$$a^{\frac{1}{n}} = \sqrt[n]{a}.$$

Problem 1   Rewrite the following expressions in the form of a^r.

(1) $\sqrt[4]{a}$ (2) $\left(\sqrt{a}\right)^3$ (3) $\sqrt[3]{a^5}$ (4) $\sqrt[3]{\dfrac{1}{a^2}}$

Problem 2   Rewrite the following expressions in the form $\sqrt[n]{a^m}$:

(1) $a^{\frac{1}{2}}$ (2) $a^{\frac{4}{3}}$ (3) $a^{-\frac{5}{3}}$ (4) $a^{1.5}$

If we define a power with a radical exponent as on the preceding page, the laws (1) - (5) of exponents still hold even when the exponents are any rational numbers.

Problem 3   Calculate the following expressions:

(1) $a^{\frac{3}{4}} \times a^{\frac{1}{2}}$ (2) $x^{\frac{1}{4}} \div x^{\frac{1}{2}}$

(3) $\left(a^{\frac{2}{3}} a^{-1}\right)^{-3}$ (4) $\left(a^{-\frac{3}{4}}\right)^{-\frac{2}{3}} \div \sqrt{a^3}$

Demonstration   Prove that $a^{\frac{1}{3}} < a^{\frac{1}{2}}$ for $a > 1$.

[Proof] $\dfrac{a^{\frac{1}{2}}}{a^{\frac{1}{3}}} = a^{\frac{1}{2} - \frac{1}{3}} = a^{\frac{1}{6}} = \sqrt[6]{a}$

If $1 < a$, then $1 < \sqrt[6]{a}$. Therefore,

$$1 < \dfrac{a^{\frac{1}{2}}}{a^{\frac{1}{3}}}.$$

Thus,

$$a^{\frac{1}{3}} < a^{\frac{1}{2}}.$$

Problem 4   Compare the values of $a^{-\frac{1}{3}}$ and $a^{-\frac{1}{2}}$ for $a > 1$.

Problem 5   Prove that $a^{1.5} > a^2$ for $0 < a < 1$.

For $a > 0$, we can define the meaning of a^x even if the exponent x is an irrational number. For example, with $\sqrt{2} = 1.4142...$, the values of

$$a^1, \ a^{1.4}, \ a^{1.41}, \ a^{1.414}, \ a^{1.4142}, \ ...$$

approach a certain value. We can designate that value as a.

In this way, even when we extend the range of exponents to encompass real numbers, the laws of exponents continue to hold.

Any power a^x of a positive number a is itself a positive number. The following relations hold for the comparison of powers.

For $a > 1$,

$$\text{if } u < v, \text{ then } a^u < a^v.$$

For $0 < a < 1$,

$$\text{if } u < v, \text{ then } a^u > a^v.$$

Exponential Functions

If a is a positive constant not equal to 1, a function which takes the form

$$y = a^x$$

is called an **exponential function** of **base** a.

Let's consider the function represented by

$$y = 2^x.$$

Problem 1   Find the values of y corresponding to the values of x in the following table. Then mark the points whose coordinates are given by (x, y) on the coordinate plane.

x	-2	-1.5	-1	-0.5	0	0.5	1	1.5	2	2.5	3
$y=2^x$											

The graph of this function is the curve in the figure below.

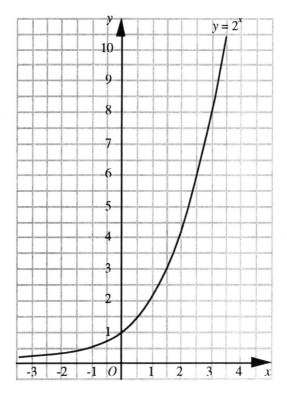

Problem 2   Compile tables for the following exponential functions as in Problem 1 for the values $-2 \le x \le 2$, and graph the functions.

(1) $y = 3^x$ (2) $y = \left(\frac{1}{2}\right)^x$ (3) $y = \left(\frac{1}{3}\right)^x$

Problem 3   What is the relation between the graphs of $y = 2^x$ and $y = \left(\frac{1}{2}\right)^x$?

Between the graphs of $y = 3^x$ and $y = \left(\frac{1}{3}\right)^x$?

In general, the graphs of $y = a^x$ and $y = \left(\frac{1}{a}\right)^x$ are symmetric with respect to the y-axis.

Problem 4   Which of the following exponential functions have graphs that are symmetric to each other with respect to the y-axis?

(1) $y = 4x$ (2) $y = \left(\frac{2}{3}\right)^x$ (3) $y = 1.5^x$

(4) $y = 0.5^x$ (5) $y = 0.25^x$

Let's summarize the properties of the graph of the exponential function $y = a^x$.

(i)   For a > 1:

If $u < v$, then $a^u < a^v$. Therefore, when x increases, a^x also increases. When x increases without bound, a^x also increases without bound, and when x is negative and the absolute value of x increases without bound, a^x approaches infinitesimally close to 0. Thus, the x-axis is the asymptote of this graph.

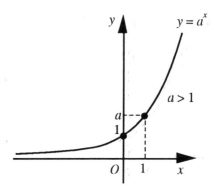

(ii)   For $0 < a < 1$:

The graphs of $y = a^x$ and $y = \left(\frac{1}{a}\right)^x$ are symmetric with respect to the y-axis. Moreover, since $\frac{1}{a} > 1$, when x increases, a^x decreases, and the graph takes the form shown in the figure to the right.

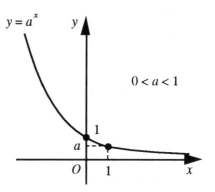

The domain of the exponential function $y = a^x$ is all real numbers, and its range is $\{y \,|\, y > 0\}$.

(**Problem 5**)   Graph the following functions based on the graph of $y = 2^x$:

(1) $y = 1 + 2^x$

(2) $y = 2^{x-1}$

Exercises

1. Find the following values:

   (1) $32^{\frac{1}{5}}$
   (2) $8^{-\frac{1}{3}}$
   (3) $25^{\frac{3}{2}}$
   (4) $(\sqrt[3]{4})^{12}$

2. Calculate the following expressions:

   (1) $\sqrt{a}\,\sqrt[3]{a}\,\sqrt[6]{a}$
   (2) $\sqrt[3]{a^2} \div \sqrt{a}$
   (3) $\dfrac{\sqrt[4]{a^3}\,\sqrt[3]{a^2}}{\sqrt[12]{a^{11}}}$

3. Calculate the following expressions:

   (1) $a^{\frac{2}{3}} \div a^{\frac{3}{2}}$
   (2) $(x^{\frac{3}{4}} \times x^{\frac{1}{2}})^{\frac{2}{5}}$
   (3) $(y^{\frac{1}{2}} + y^{-\frac{1}{2}})^2$

4. Find the values of x that satisfy the following equalities:

   (1) $4^x = 32$
   (2) $(\frac{1}{27})^x = 81$
   (3) $125^x = \frac{1}{25}$

5. Find the value of $\dfrac{2^{3x} + 2^{-3x}}{2^x + 2^{-x}}$ for $2^x = 5$.

6. Arrange the following numbers from largest to smallest:

   $$\sqrt{2} \qquad \sqrt[3]{2} \qquad \sqrt[5]{4} \qquad \sqrt[8]{8} \qquad \sqrt[9]{16}$$

7. What is the relation between the following graphs and the graph of $y = 2^x$?

   (1) $y = -2^x$
   (2) $y = \dfrac{1}{2^x}$
   (3) $y = -2^{-x}$

8. Graph $y = 2^{-x+2}$ based on the graph of $y = 2^x$.

LOGARITHMIC FUNCTIONS

Logarithms

For $a > 0$ and $a \neq 1$, you can see from the graph of the exponential function $y = a^x$ that a real number q satisfying

$$p = a^q$$

takes on a single value for any positive number p. This q can be expressed as

$$\log_a p \; *$$

and called the **logarithm** of p to the **base** a. Moreover, p is called the **antilogarithm** of q to the base a.

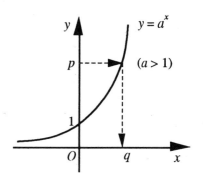

Definition of a Logarithm

$$p = a^q \iff q = \log_a p$$

Note: When we write $\log_a p$, it should be stipulated that $a > 0$, $a \neq 1$, and $p > 0$. Henceforth we will not specify this condition every time.

Example $8 = 2^3 \iff 3 = \log_2 8$

$100 = 10^2 \iff 2 = \log_{10} 100$

---

* log is an abbreviation of logarithm.

Problem 1   Rewrite the following equalities in the form $q = \log_a p$:

(1) $125 = 5^3$

(2) $2^{-2} = 0.25$

(3) $8^{-\frac{1}{3}} = 0.5$

Problem 2   Rewrite the following equalities in the form $p = a^q$:

(1) $\frac{3}{2} = \log_4 8$

(2) $-\frac{1}{2} = \log_{64} \frac{1}{8}$

(3) $0 = \log_{10} 1$

Demonstration   Find the value of $\log_8 4$.

[Solution]   If we set $\log_8 4 = x$, then

$$8^x = 4.$$

Since $8^x = (2^3)^x = 2^{3x}$ and $4 = 2^2$,

$$2^{3x} = 2^2.$$

Therefore,

$$3x = 2.$$

Thus, $$x = \frac{2}{3}.$$

Answer: $\log_8 4 = \frac{2}{3}$

Problem 3   Find the following values:

(1) $\log_3 27$

(2) $\log_2 \sqrt{2}$

(3) $\log_7 1$

(4) $\log_5 5$

(5) $\log_{10} \sqrt[3]{0.01}$

(6) $\log_{\sqrt{3}} 3^{-1}$

Properties of Logarithms

Let's examine various properties of logarithms.

First, from $a^0 = 1$ and $a^1 = a$, we can see that:

$$\log_a 1 = 0; \qquad\qquad \log_a a = 1.$$

Moreover, from laws (1) - (3) of exponents on page 8, we can deduce the following formulas for positive numbers M and N.

Properties of Logarithms

[I] $\log_a MN = \log_a M + \log_a N$ Logarithm of a product

[II] $\log_a \dfrac{M}{N} = \log_a M - \log_a N$ Logarithm of a quotient

[III] $\log_a M^r = r \log_a M$ Logarithm of a power

Let's prove property [I].

If we set $\log_a M = x$ and $\log_a N = y$, then

$$M = a^x; \qquad\qquad N = a^y.$$

Therefore,

$$MN = a^x \times a^y = a^{x+y}.$$

Thus,

$$\log_a MN = x + y$$

$$= \log_a M + \log_a N.$$

Problem 1   Prove properties [II] and [III] in the same way we proved [I].

Problem 2   Prove that the following equalities hold:

(1) $\log_a \dfrac{1}{N} = -\log_a N$ (2) $\log_a \sqrt[n]{M} = \dfrac{1}{n} \log_a M$

(Demonstration 1)   Simplify $\log_2 \frac{4}{3} + \log_2 24$.

[Solution] $\log_2 \frac{4}{3} + \log_2 24 = \log_2 \left(\frac{4}{3} \times 24 \right)$

$$= \log_2 32$$

$$= \log_2 2^5 = 5$$

(Problem 3)   Simplify the following expressions:

    (1) $\log_6 \frac{9}{2} + \log_6 8$ (2) $\log_5 250 - \log_5 2$

    (3) $\frac{1}{2} \log_2 25 - \log_2 10$ (4) $\dfrac{\log_3 32}{\log_3 8}$

(Problem 4)   Set $\log_{10} 2 = a$ and $\log_{10} 3 = b$. Express the value of the following expressions using a and b:

    (1) $\log_{10} 6$ (2) $\log_{10} 5$ (3) $\log_{10} \sqrt{18}$

    (4) $\log_{10} 120$ (5) $\log_{10} \sqrt{0.2}$

    A logarithm to the base 10 is called a **common logarithm**, and we have tables of common logarithms to show their values. The Appendix includes a table of common logarithms for four-digit numbers.

    The value of a logarithm to a base other than 10 can be calculated using the following formula.

Base-Changing Formula

$$\log_a b = \frac{\log_c b}{\log_c a}$$

[Proof]   If we set $\log_a b = x$, then $a^x = b$

Take the logarithm to the base c of both sides.

$$\log_c a^x \ = \ \log_c b$$

Therefore,
$$x \log_c a \ = \ \log_c b.$$

Since $a \neq 1$,
$$\log_c a \ \neq \ 0.$$

Thus,
$$x \ = \ \frac{\log_c b}{\log_c a} \ .$$

Therefore,
$$\log_a b \ = \ \frac{\log_c b}{\log_c a} \ .$$

Problem 5   Prove the following equalities:

(1) $\log_a b = \dfrac{1}{\log_b a}$ (2) $\log_a b \ \log_b c \ \log_c a = 1$

Problem 6   Find the values of the following expressions, taking $\log_{10} 2 = 0.3010$ and $\log_{10} 3 = 0.4771$:

(1) $\log_2 3$ (2) $\log_2 10$ (3) $\log_3 20$

Demonstration 2   Find the value of x which satisfies $2^x = 5$ from the table of common logarithms.

[Solution]   Take the common logarithm of both sides.

$$\log_{10} 2^x \ = \ \log_{10} 5$$

Therefore,
$$x \log_{10} 2 \ = \ \log_{10} 5.$$

From the table of common logarithms,
$$0.3010x \ = \ 0.6990 \ .$$

Thus,
$$x \ = \ \frac{0.6990}{0.3010} \approx 2.32 \ .$$

(**Problem 7**) Find the values of x which satisfy the following equalities by the same method as in Demonstration 2:

(1) $3^x = 6$ (2) $2^x = 100$

(**Demonstration 3**) ^{237}U (uranium 237) decays at a fixed rate such that after 7 days half of the original amount remains. How many days will it take until $\dfrac{1}{10}$ of the original amount of uranium remains? Take $\log_{10} 2 = 0.3010$.

[Solution] If we assume the amount remaining after one day will be a times the original amount, then

$$a^7 = \frac{1}{2} .$$

Therefore,

$$a = \left(\frac{1}{2} \right)^{\frac{1}{7}} = 2^{-\frac{1}{7}}.$$

If we assume that it takes n days for the uranium to be reduced to $\dfrac{1}{10}$ of the original amount, then

$$\left(2^{-\frac{1}{7}} \right)^n = \frac{1}{10} .$$

Take the common logarithm of both sides:

$$\log_{10} \left(2^{-\frac{1}{7}} \right)^n = \log_{10} \frac{1}{10} ,$$

$$-\frac{n}{7} \log_{10} 2 = -1.$$

Therefore,

$$n = \frac{7}{\log_{10} 2} = \frac{7}{0.3010} = 23.2... .$$

Answer: 24 days

Problem 8 When light passes through one pane of a certain kind of glass, the light loses 10% of its intensity. How many panes of glass would it take for light to be reduced to $\frac{1}{3}$ or less of its original intensity? Here, we assume $\log_{10} 3 = 0.4771$.

Problem 9 A certain type of bacteria doubles in number every 20 minutes. How long will it take for 20 bacteria to increase to more than 1 million? Take $\log_{10} 2 = 0.3010$.

 Logarithmic Functions

If y , a function of x , is expressed as

$$y = \log_a x,$$

this function is called a **logarithmic function** of x to the base a.

You can see from the definition of a logarithm on page 15 that the logarithmic function $y = \log_a x$ is the inverse of the exponential function $y = a^x$. Therefore, the domain of $y = \log_a x$ is $\{x \mid x > 0\}$ and the range is the set of all real numbers.

Graphs of Logarithmic Functions

The graphs of two functions which are inverse to each other are symmetric with respect to the straight line $y = x$. Therefore, if we fold a graph of the exponential function $y = a^x$ with the straight line $y = x$ as the axis, we obtain the graph of the logarithmic function $y = \log_a x$.

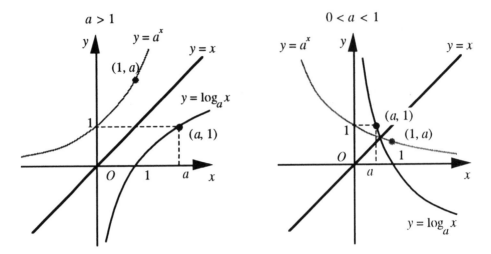

$a > 1$ $0 < a < 1$

Problem 1   Graph $y = \log_2 x$ and $y = \log_{\frac{1}{2}} x$ using the graphs of $y = 2^x$ and

$$y = \left(\tfrac{1}{2}\right)^x.$$

We can infer the following facts pertaining to the logarithmic function $y = \log_a x$ from the properties of exponential functions and from the above figure:

(1)   The graph exists on the interval $x > 0$.

(2)   The graph passes through the point $(1, 0)$.

(3)   The y-axis is the asymptote of the graph.

(4)   For $a > 1$,

as x increases, $\log_a x$ increases.

For $0 < a < 1$,

as x increases, $\log_a x$ decreases.

Demonstration 1   Find the range of the value of $\log_{10} x$ for $1 < x < 10$.

[Solution]   Take the common logarithm of each term of $1 < x < 10$.

Since the base is greater than 1,

$$\log_{10} 1 < \log_{10} x < \log_{10} 10.$$

Thus,

$$0 < \log_{10} x < 1.$$

Problem 2   Find the range of the value of $\log_{10} x$ when x ranges on the following values:

(1) $10 < x < 100$ (2) $0.01 < x \leq 1$

(3) $0.001 \leq x \leq 1{,}000$ (4) $10^n \leq x < 10^{n+1}$

Demonstration 2   How many digits are there in 2^{30}? Take $\log_{10} 2 = 0.3010$.

[Solution]   Set $x = 2^{30}$, and take the common logarithm of both sides.

$$\log_{10} x = \log_{10} 2^{30}$$

$$= 30 \log_{10} 2 = 30 \times 0.3010 = 9.03$$

Therefore,
$$9 < \log_{10} x < 10.$$

Since $9 = \log_{10} 10^9$ and $10 = \log_{10} 10^{10}$,

$$10^9 < x < 10^{10}.$$

Thus, x has 10 digits.

Problem 3   How many digits are there in 3^{50}? Take $\log_{10} 3 = 0.4771$.

Exercises

1. Fill in the blanks with appropriate numbers:

   (1) $\log_3 \boxed{} = -2$ (2) $\log_{\boxed{}} 8 = 3$

   (3) $\log_{10} 0.01 = \boxed{}$ (4) $\log_a \sqrt{a} = \boxed{}$

   (5) $\log_7 \boxed{} = 2$ (6) $\log_5 10 = \dfrac{1}{\log_{10} \boxed{}}$

2. Find the value of the following expressions:

   (1) $(\log_2 3)(\log_3 16)$

   (2) $(\log_2 6)(\log_3 6) - (\log_2 3 + \log_3 2)$

   (3) $(\log_{10} 2)^2 + (\log_{10} 5)(\log_{10} 4) + (\log_{10} 5)^2$

3. What is the relation between the graph of $y = \log_3 x$ and the graphs of the following functions?

   (1) $y = \log_3 \dfrac{1}{x}$ (2) $y = \log_{\frac{1}{3}} x$ (3) $y = \log_{\frac{1}{3}} \dfrac{1}{x}$

4. Compare the values of each pair of expressions:

   (1) $3 \log_4 3,\ 2 \log_2 3$ (2) $\log_3 2,\ \dfrac{2}{3}$

5. Compare the value of 2^{30} and 3^{20}, taking $\log_{10} 2 = 0.3010$ and $\log_{10} 3 = 0.4771$.

6. Answer the following questions, taking $\log_{10} 2 = 0.3010$:

   (1) How many digits are there in 4^{15}?

   (2) At what decimal place in $\left(\dfrac{1}{4}\right)^{20}$ does the first number other than 0 appear?

7.   Find the inverse of the following functions:

   (1) $y = 3^{x-1}$ (2) $y = 2\log_4 x$

8.   Prove that the following equalities hold:

   (1) $\log_a b = \log_{a^2} b^2 = \log_{a^3} b^3$ (2) $a = b^{\log_b a}$

Chapter Exercises

A

1.   Find the following values:

   (1) $16^{\frac{3}{2}}$ (2) $8^{-\frac{2}{3}}$

   (3) $0.25^{-\frac{1}{2}}$ (4) $4^{3.5}$

2.   Perform the following calculations:

   (1) $\left(ab^2\right)^{\frac{2}{3}} \times a^{\frac{1}{2}} \div b^{\frac{1}{3}}$ (2) $\left(x^{\frac{1}{2}} y^{-\frac{1}{2}}\right)\left(x^{-1} y\right)$

3.   Find the value of the following expressions if $x^{\frac{1}{2}} + x^{-\frac{1}{2}} = 4$:

   (1) $x + x^{-1}$ (2) $x^2 + x^{-2}$

4.   Express y as a power of x, if $a^{\frac{1}{2}} = x$ and $a^{\frac{2}{3}} = y$.

5.   Find the following values:

(1) $\log_4 0.5$ (2) $\log_2 \frac{1}{32}$

(3) $\log_6 24 + \log_6 3 - \log_6 2$

6.   Express the values of the following expressions in terms of m and n, if $\log_a x = m$ and $\log_a y = n$.

(1) $\log_a x^3 + \log_a \sqrt{y}$ (2) $\log_a \sqrt{x} - \log_a y^2$

(3) $\log_a a^2 xy$ (4) $\log_a xy^3$

7.   Find the values of x which satisfy the following equalities:

(1) $\log_x 8 = \frac{3}{2}$ (2) $\log_3 x = -3$ (3) $\log_{10} x^2 = 4$

8.   Demonstrate that the following formulas hold for $f(x) = a^x$.

(1) $f(x)f(y) = f(x + y)$ (2) $\dfrac{f(x)}{f(y)} = f(x - y)$

9.   Demonstrate that the following formulas hold for $f(x) = \log_a x$.

(1) $f(xy) = f(x) + f(y)$ (2) $f\left(\dfrac{x}{y}\right) = f(x) - f(y)$

10.   Graph the following functions:

(1) $y = 2^{x+1}$ (2) $y = \log_2 x - 1$

𝔹

1.   Simplify the following expressions:

(1) $\left(a^{\frac{1}{3}} - a^{-\frac{1}{3}}\right)\left(a^{\frac{2}{3}} + 1 + a^{-\frac{2}{3}}\right)$ (2) $(x^a)^{b-c}(x^b)^{c-a}(x^c)^{a-b}$

2.   Find the values of the following expressions, if $a^{2x} = 3$.

(1) $(a^x + a^{-x})^2$ (2) $\dfrac{a^{3x} + a^{-3x}}{a^x + a^{-x}}$

3.   Find the value of $2x^3 + 6x$, if $x = 2^{\frac{1}{3}} - 2^{-\frac{1}{3}}$.

4.   Find the value of x which satisfies the equality $9^x + 3^x = 12$, taking $3^x = t$.

5. a, b, and c are positive numbers. If the following statements are true, then prove them, and if they are false, find a counterexample.

(1)   For $a < b$, $a^c < b^c$ holds.

(2)   For $a < b$, $c^a < c^b$ holds.

6.   At what decimal place does the first non-zero digit occur if the following numbers are expressed in decimal form? Assume that $\log_{10} 2 = 0.3010$ and $\log_{10} 3 = 0.4771$.

(1) $\left(\frac{1}{5}\right)^{15}$ (2) $\left(\frac{2}{3}\right)^{10}$

7.   Find the minimum natural number n which satisfies each of the following inequalities. Take $\log_{10} 2 = 0.3010$ and $\log_{10} 3 = 0.4771$.

(1) $1.5^n > 100$ (2) $0.4^n < 0.001$

8.   Find the range of values of x which satisfy the following inequalities:

(1) $\log_3 (x - 3) \le 3$ (2) $\log_{0.5} 2x \le 2$

9.   Find the value of x which satisfies $(\log_{10} x)^2 - \log_{10} x = 2$ by setting $\log_{10} x = t$.

CHAPTER 2

TRIGONOMETRIC FUNCTIONS

SECTION 1.   TRIGONOMETRIC FUNCTIONS
SECTION 2.   ADDITION THEOREMS

    It was only in the modern period, when change with respect to time was recognized as an object of mathematical investigation, that a new meaning was found for trigonometric functions with real number variables, based on the trigonometric ratios that had served in measurement, astronomy, and navigation since ancient times. One of the most remarkable phenomena recognized as a type of change with respect to time was periodic motion, such as the movement of a planet or the oscillation of a string. It is obvious that trigonometric functions, which are the most fundamental periodic functions, played a very important role in describing periodic motion.

    Subsequently, trigonometric functions gave rise to the problem of representing an arbitrary function as an infinite sum of trigonometric functions. Trigonometric functions acted as a midwife in the birth of the modern concept of the function. Moreover, trigonometric functions assumed the principal role in mathematical physics, which led to modern quantum physics through the so-called "superposition principle."

TRIGONOMETRIC FUNCTIONS

 General Angles

Let's rotate a ray OP in a plane, taking point O as the center.

There are two directions in which we can rotate it. We define the **positive direction** as counterclockwise and the **negative direction** as clockwise.

For example, if ray OP starts from ray OX and rotates 120° in the positive direction, and then rotates 155° in the negative direction, OP rotates a total of 35° in the negative direction from the original position.

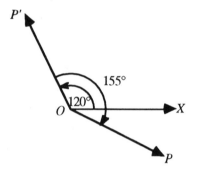

A rotating ray such as OP is called the **terminal side**, while the ray that indicates the base position for the rotation, here ray OX, is called the **initial side**.

Problem 1   Terminal side OP rotates about center O as indicated below. What is the total rotation of OP from the initial side in terms of direction and degrees?

(1)   70° in the positive direction, and then 150° in the positive direction.

(2)   135° in the positive direction, and then 180° in the negative direction.

(3)   60° in the negative direction, and then 300° in the positive direction.

Rotation of terminal side OP 30° in the positive direction about center O can be referred to simply as rotation of 30°, while the rotation of 30° in the negative direction is rotation of -30°.

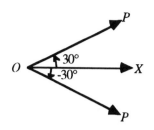

Example 1   If terminal side OP rotates 120° from initial side OX, and then rotates -155°, the total rotation of OP from the initial side is

$$120° + (-155°) = -35°.$$

Example 2   If terminal side OP rotates 125° from initial side OX, and then rotates 270°, the total rotation of OP from the initial side is

$$125° + 270° = 395°.$$

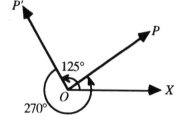

Problem 2   How many degrees do the minute hand and the hour hand of a clock rotate between 8:00 pm and 10:00 am?

As illustrated by the above Examples and Problem, we sometimes need to regard an angle greater than 360° or a negative angle as angles of rotation. An angle with this broad interpretation is called a **general angle**. A ray specified by rotation of α about center O from the position of initial side OX is called the **terminal side of** α. If we rotate the terminal side of α by β, it will become the terminal side of α + β.

Problem 3   What rotation of the terminal side of α gives us the terminal side of α − β?

Problem 4   Mark the terminal sides of 290°, 800°, and -585° on a figure.

Let us define initial side OX and take OP as the terminal side of α. Now we see from the figure to the right that the following angles also have the same terminal side OP.

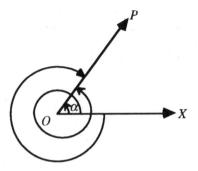

$\alpha + 360°$ $\alpha - 360°$

$\alpha + 2 \times 360°$ $\alpha - 2 \times 360°$

$\alpha + 3 \times 360°$

All such angles can be referred to as the **angles belonging to terminal side** OP.

> The angles belonging to terminal side OP of α are
>
> $\alpha + n \times 360°$ (where n is an integer).

(Problem 5)   In the following figures, state the angles belonging to terminal side OP with ray OX as the initial side:

(1)   (2)   (3)

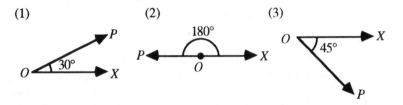

Radian Measure

We are very familiar with the sexagesimal (base 60) system for measuring an angle: we define $\frac{1}{90}$ of a right angle as $1°$ (1 degree), $\frac{1}{60}$ of $1°$ as $1'$ (1 minute), and $\frac{1}{60}$ of $1'$ as $1''$ (1 second).

However, there is also another system for measuring an angle: in a single circle, we take as the unit of measurement the central angle subtended by an arc whose length is equal to the radius of the circle.

Let's express central angle α by means of the sexagesimal system.

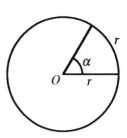

The length of the arc is proportional to the central angle. Therefore, if we take r as the radius of the circle, then

$$r : 2\pi r = \alpha : 360°.$$

Thus, $\qquad \alpha = \dfrac{180°}{\pi} \approx 57.2958°.$

This central angle is a constant angle unrelated to the radius of the circle. The measure of this angle is said to be **1 radian**, and the system of measuring angles using this angle as the basic unit is called **radian measure**.

$$1 \text{ radian} = \frac{180°}{\pi}, \quad 1° = \frac{\pi}{180} \text{ radians}$$

If an angle of $x°$ in the sexagesimal system is y radians, then

$$y = \frac{\pi}{180}x.$$

When an angle is expressed in radian measure, the unit name "radians" is usually omitted.

Example 3 $\quad 180° = \pi, \qquad\qquad 90° = \dfrac{\pi}{2}, \qquad\qquad 60° = \dfrac{\pi}{3}$

Problem 6 $\quad$ Express 30°, 45°, 150°, and 360° in radian measure.

Problem 7 $\quad$ Express radian measures $\dfrac{\pi}{5}, \dfrac{2}{3}\pi, \dfrac{3}{2}\pi$, and 3π in the sexagesimal system.

Problem 8 $\quad$ Show the terminal sides of the following angles in radian measure by means of a diagram:

(1) $\dfrac{10}{3}\pi$ $\qquad\qquad$ (2) $-\dfrac{13}{6}\pi$ $\qquad\qquad$ (3) $\dfrac{9}{2}\pi$

Any angle belonging to terminal side OP can be expressed in terms of an integer n and θ , some angle belonging to that terminal side:

$$\theta + 2n\pi \qquad \text{(where } n \text{ is an integer).}$$

The angle θ is usually taken to fall in the range $0 \le \theta < 2\pi$ or $-\pi < \theta \le \pi$.

Problem 9   The following angles are given as belonging to terminal side OP . Express the general angle belonging to terminal side OP in radian measure.

(1) $\dfrac{\pi}{3}$ (2) $-\dfrac{3}{4}\pi$ (3) 3π

From now on, we will use radian measure to express angles.

 Trigonometric Functions

You have already studied the trigonometric ratios of the angles between 0° and 180° in Mathematics I. Here let's define the sine, cosine, and tangent of a general angle.

On the coordinate plane consider the angle belonging to a terminal side rotated about the origin O , with the positive ray of the x -axis as the initial side.

Draw the circle O with a radius of r and its center at the origin. Take point P as the intersection of the terminal side of θ and circle O . When the coordinates of P are (x, y) , the values of the ratios

$$\frac{y}{r}, \ \frac{x}{r}, \ \frac{y}{x}$$

depend on the angle θ , and not on the radius r of circle O .

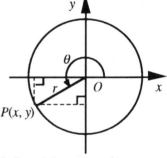

These values are called **trigonometric functions** of θ , and they are written as

$$\sin \theta = \frac{y}{r}, \qquad \cos \theta = \frac{x}{r}, \qquad \tan \theta = \frac{y}{x}. \tag{1}$$

Note that $\tan \theta$ cannot be defined for θ such that $x = 0$.

If θ is an angle in the range $0 \le \theta \le \pi$, the values of the trigonometric functions by the above definition are identical to the values you learned in Mathematics I.

Problem 1 Find the sine, cosine, and tangent of $\dfrac{\pi}{3}$, $\dfrac{\pi}{4}$, and $\dfrac{\pi}{6}$.

The following formula holds based on (1) above.

[I] $\tan \theta = \dfrac{\sin \theta}{\cos \theta}$

Problem 2 Check that formula [I] holds.

A circle with its center at the origin and a radius of 1 is called the **unit circle**.

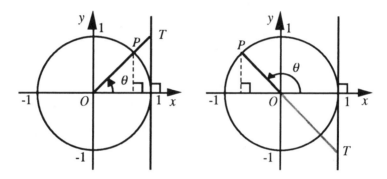

As in the figure above, take P as the intersection of the unit circle and the terminal side of angle θ, and take T as the intersection of straight line OP and the line $x = 1$, the tangent line of the unit circle at the point $(1, 0)$. The coordinates of P and T are, respectively,

$$(\cos \theta, \sin \theta), \qquad (1, \tan \theta).$$

Problem 3 Check that the coordinates of P and T given above are correct.

Since the point $P(\cos \theta, \sin \theta)$ lies on the circumference of the circle $x^2 + y^2 = 1$, the following generalizations can be made:

[II] $\sin^2 \theta + \cos^2 \theta = 1$;

[III] $-1 \le \sin \theta \le 1$, $\qquad -1 \le \cos \theta \le 1$.

Demonstration 1   Find the sine, cosine, and tangent of $-\dfrac{\pi}{4}$.

[Solution]   If we take point P as in the figure to the right, then the coordinates of P are

$$\left(\frac{1}{\sqrt{2}},\ -\frac{1}{\sqrt{2}} \right).$$

Therefore,

$$\sin \left(-\frac{\pi}{4} \right) = -\frac{1}{\sqrt{2}},$$

$$\cos \left(-\frac{\pi}{4} \right) = \frac{1}{\sqrt{2}},$$

$$\tan \left(-\frac{\pi}{4} \right) = -1.$$

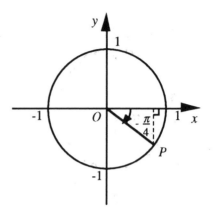

In Demonstration 1, the terminal side of $-\dfrac{\pi}{4}$ lies in the fourth quadrant. When the terminal side of angle θ lies in the fourth quadrant in this way, we say that angle θ is "an angle in the fourth quadrant". We speak of angles in the other quadrants in the same way.

The sign of a trigonometric function is fixed, based on the quadrant; the possibilities are given in the table to the right.

Quadrant	1	2	3	4
$\sin \theta$	+	+	–	–
$\cos \theta$	+	–	–	+
$\tan \theta$	+	–	+	–

Problem 4   Check the table to the right.

Problem 5   Find the following values:

(1) $\cos (-60°)$ (2) $\sin (-135°)$ (3) $\tan 870°$

(4) $\sin \dfrac{4}{3} \pi$ (5) $\tan \left(-\dfrac{5}{4} \pi \right)$ (6) $\cos \left(-\dfrac{7}{2} \pi \right)$

Problem 6   Prove that the following equalities hold:

(1) $1 + \tan^2 \theta = \dfrac{1}{\cos^2 \theta}$

(2) $\dfrac{1 - 2\sin \theta \cos \theta}{1 + 2\sin \theta \cos \theta} = \left(\dfrac{1 + \tan \theta}{1 + \tan \theta}\right)^2$

(3) $\sin^4 \theta - \cos^4 \theta = 2\sin^2 \theta - 1$

Demonstration 2   Find the value of $\cos \theta$ and $\tan \theta$, if $\sin \theta = -\dfrac{3}{5}$ and θ is an angle in the third quadrant.

[**Solution**]   Since $\sin^2 \theta + \cos^2 \theta = 1$ and $\sin \theta = -\dfrac{3}{5}$,

$$\cos \theta = \pm \sqrt{1 - \left(-\dfrac{3}{5}\right)^2} = \pm \dfrac{4}{5}.$$

Since θ is an angle in the third quadrant,

$$\cos \theta = -\dfrac{4}{5}.$$

From $\tan \theta = \dfrac{\sin \theta}{\cos \theta}$,

$$\tan \theta = \dfrac{3}{4}.$$

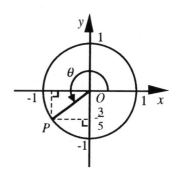

Problem 7   Find the answers as in Demonstration 2 if θ is an angle in the fourth quadrant.

Problem 8   Find the values of $\sin \theta$ and $\tan \theta$, if $\cos \theta = -\dfrac{5}{13}$.

Properties of Trigonometric Functions

Since the angles θ and $\theta + 2n\pi$ have the same terminal side, the following formulas hold, provided that n is an integer.

[IV] $\sin(\theta + 2n\pi) = \sin\theta$

$\cos(\theta + 2n\pi) = \cos\theta$

$\tan(\theta + 2n\pi) = \tan\theta$

Example 1 $\sin\dfrac{9}{2}\pi = \sin\left(\dfrac{\pi}{2} + 4\pi\right) = \sin\dfrac{\pi}{2} = 1$

$\cos\left(-\dfrac{17}{3}\pi\right) = \cos\left(\dfrac{\pi}{3} - 6\pi\right) = \cos\dfrac{\pi}{3} = \dfrac{1}{2}$

Problem 1   Find the values of $\sin 6\pi$, $\sin\dfrac{11}{3}\pi$, $\cos\left(-\dfrac{23}{6}\pi\right)$, and $\tan\left(-\dfrac{27}{4}\pi\right)$.

The following formulas relate the trigonometric functions of θ and $-\theta$.

[V] $\sin(-\theta) = -\sin\theta$ $\cos(-\theta) = \cos\theta$

$\tan(-\theta) = -\tan\theta$

[Proof]   The coordinates of points P and P', where the terminal sides of θ and $-\theta$ intersect the unit circle, are

$$(\cos\theta, \sin\theta), \quad (\cos(-\theta), \sin(-\theta)).$$

Since P and P' are symmetric with respect to the x-axis,

$$\sin(-\theta) = -\sin\theta,$$

$$\cos(-\theta) = \cos\theta.$$

Therefore,

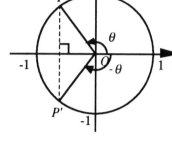

$$\tan(-\theta) = \frac{\sin(-\theta)}{\cos(-\theta)}$$

$$= \frac{-\sin\theta}{\cos\theta}$$

$$= -\tan\theta.$$

Example 2 $\sin\left(-\dfrac{\pi}{3}\right) = -\sin\dfrac{\pi}{3} = -\dfrac{\sqrt{3}}{2}$

$\cos\left(-\dfrac{\pi}{3}\right) = \cos\dfrac{\pi}{3} = \dfrac{1}{2}$

The following formulas relate the trigonometric functions of θ and $\dfrac{\pi}{2} - \theta$.

[VI] $\sin\left(\dfrac{\pi}{2} - \theta\right) = \cos\theta$ $\cos\left(\dfrac{\pi}{2} - \theta\right) = \sin\theta$

$\tan\left(\dfrac{\pi}{2} - \theta\right) = \dfrac{1}{\tan\theta}$

[Proof]　If we take P as the point where the terminal side of θ intersects the unit circle, then the coordinates of P are

$$(\cos \theta, \sin \theta).$$

If we take P' as the point symmetric to P with respect to the straight line $y = x$, then the coordinates of P' are

$$(\sin \theta, \cos \theta).$$

However, as you can see from the figure to the right, ray OP' is the terminal side of $\frac{\pi}{2} - \theta$.

Therefore, the coordinates of P' are

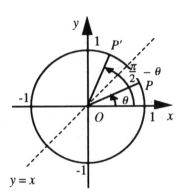

$$\left(\cos\left(\frac{\pi}{2} - \theta\right), \sin\left(\frac{\pi}{2} - \theta\right)\right).$$

Thus,

$$\sin\left(\frac{\pi}{2} - \theta\right) = \cos \theta, \qquad \cos\left(\frac{\pi}{2} - \theta\right) = \sin \theta.$$

Moreover,

$$\tan\left(\frac{\pi}{2} - \theta\right) = \frac{\sin\left(\frac{\pi}{2} - \theta\right)}{\cos\left(\frac{\pi}{2} - \theta\right)} = \frac{\cos \theta}{\sin \theta} = \frac{1}{\tan \theta}.$$

The following formulas relate the trigonometric functions of θ and $\pi - \theta$.

[VII]　$\sin(\pi - \theta) = \sin \theta$　　　$\cos(\pi - \theta) = -\cos \theta$

$\tan(\pi - \theta) = -\tan \theta$

Problem 2 Prove formula [VII] by referring to the figure at the right.

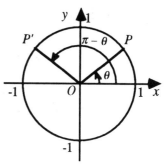

The following formulas are obtained by substituting θ for $-\theta$ in formulas [VI] and [VII], and then applying formula [V].

[VI] $\sin (\theta + \dfrac{\pi}{2}) = \cos \theta \quad \cos (\theta + \dfrac{\pi}{2}) = - \sin \theta$

$\tan (\theta + \dfrac{\pi}{2}) = - \dfrac{1}{\tan \theta}$

[VII] $\sin (\theta + \pi) = - \sin \theta \quad \cos (\theta + \pi) = - \cos \theta$

$\tan (\theta + \pi) = \tan \theta$

We can find the values of a trigonometric function of any general angle by converting it into a trigonometric function of an acute angle by means of these formulas.

Demonstration 1 Find the values of $\sin \dfrac{29}{6} \pi$ and $\cos (-\dfrac{20}{3} \pi)$.

[Solution] $\sin \dfrac{29}{6} \pi = \sin (\dfrac{5}{6} \pi + 4\pi) = \sin \dfrac{5}{6} \pi$

$= \sin (\pi - \dfrac{\pi}{6}) = \sin \dfrac{\pi}{6} = \dfrac{1}{2}$

$\cos (-\dfrac{20}{3} \pi) = \cos \dfrac{20}{3} \pi = \cos (\dfrac{2}{3} \pi + 6\pi)$

$= \cos \dfrac{2}{3} \pi = \cos (\pi - \dfrac{\pi}{3})$

$= - \cos \dfrac{\pi}{3} = - \dfrac{1}{2}$

Problem 3 Find the sine, cosine, and tangent of $\dfrac{17}{6} \pi$, $-\dfrac{14}{3} \pi$, and $\dfrac{15}{4} \pi$.

Demonstration 2   Find sin 1,000° using the table of trigonometric functions in the Appendix.

[Solution]   sin 1,000° = sin (360° x 3 – 80°)

$\qquad\qquad\qquad$ = sin (- 80°)

$\qquad\qquad\qquad$ = -sin 80° = - 0.9848

Problem 4   Find the values of cos 850°, sin 1,315°, and tan (-3,430°).

Demonstration 3   Find angle θ such that $\sin \theta = \dfrac{1}{2}$.

[Solution]   The y-coordinate of the point at which the terminal side of the angle θ we want to find intersects unit circle O is $\dfrac{1}{2}$.

Therefore, if we take P and P' as the points at which the straight line $y = \dfrac{1}{2}$ intersects unit circle O, then the angles belonging to terminal sides OP and OP' are the angles we want to find.

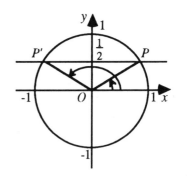

Thus, in the range $0 \le \theta < 2\pi$,

$$\theta = \frac{\pi}{6}, \ \frac{5}{6}\pi.$$

Therefore, taking n as any arbitrary integer, angle θ takes the value:

$$\theta = \frac{\pi}{6} + 2n\pi, \qquad \theta = \frac{5}{6}\pi + 2n\pi.$$

Problem 5 Find angle θ to satisfy the following equalities:

(1) $\sin \theta = \dfrac{\sqrt{3}}{2}$

(2) $\cos \theta = \dfrac{1}{2}$

(3) $\sin 2\theta = -\dfrac{1}{\sqrt{2}}$

(4) $2\cos 2\theta = 1$

Demonstration 4 Find the range of angle θ which satisfies the inequality $\cos \theta > \dfrac{\sqrt{3}}{2}$. We can assume that $0 \le \theta < 2\pi$.

[Solution] The straight line $x = \dfrac{\sqrt{3}}{2}$ intersects unit circle O at two points P and P'. The terminal side of angle θ lies inside $\angle POP'$ in the figure to the right.

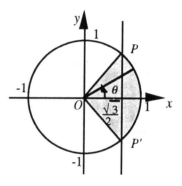

The angles belonging to OP and OP' between 0 and 2π are $\dfrac{\pi}{6}$ and $\dfrac{11}{6}\pi$, respectively.

Therefore, the range we want to find is

$$0 \le \theta < \frac{\pi}{6}, \qquad \frac{11}{6}\pi < \theta < 2\pi.$$

Problem 6 Find the range of angle θ which satisfies the following inequalities. We can assume $0 \le \theta < 2\pi$.

(1) $\cos \theta \le \dfrac{\sqrt{3}}{2}$

(2) $\sin \theta > \dfrac{1}{2}$

The Graphs of Trigonometric Functions

The Graph of $\sin \theta$

Let us take P as the intersection of the unit circle and the terminal side of angle θ, as in the figure to the right, and take (x, y) as its coordinates. Then we have

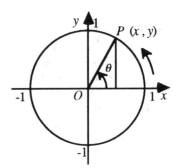

$$\sin \theta = y.$$

Thus, the value of $\sin \theta$ is equal to the y-coordinate of P.

When θ changes from 0 to $\dfrac{\pi}{2}$, P moves from $(1, 0)$ to $(0, 1)$. Therefore, the value of $\sin \theta$ increases from 0 to 1.

Analogously, when θ changes from $\dfrac{\pi}{2}$ to 2π, the value of $\sin \theta$ changes as shown in the table below:

θ	0	...	$\dfrac{\pi}{2}$	...	π	...	$\dfrac{3}{2}\pi$	...	2π
Change in $\sin \theta$	0	↗ incr.	1	↘ decr.	0	↘ decr.	-1	↗ incr.	0

incr. = increasing
decr. = decreasing

When θ is greater than 2π or less than 0, the variation described above is repeated.

When we draw the graph of $\sin \theta$, it takes the form in the figure below. This graph is called a **sine curve**.

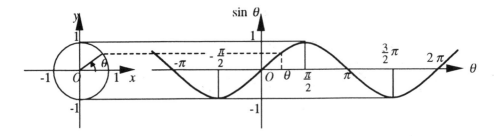

The Graph of $\cos \theta$

The figure on the preceding page shows that the value of $\cos \theta$ is equal to the x-coordinate of P. If we use this fact to draw the graph of $\cos \theta$, it takes the form in the figure below.

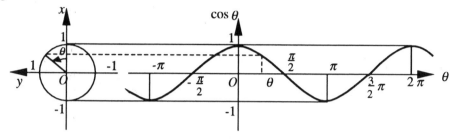

Note: The graph of $\cos \theta$ is identical to the graph created by translating the graph of $\sin \theta$ by $-\dfrac{\pi}{2}$ along the horizontal axis.

The Graph of $\tan \theta$

If we take $T(1, t)$ as the point at which the straight line defined by angle θ intersects the line $x = 1$, then

$$\tan \theta = t.$$

Thus, the value of $\tan \theta$ is equal to the y-coordinate of T.

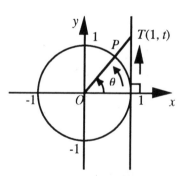

When θ changes from 0 to $\dfrac{\pi}{2}$, T moves upward. Therefore, the value of $\tan \theta$ increases. When θ changes from $\dfrac{\pi}{2}$ to π, T also moves upward in the fourth quadrant, as in the lower figure to the right.

When θ approaches $\dfrac{\pi}{2}$ from the side on which it is less than $\dfrac{\pi}{2}$, $\tan \theta$ is positive and has an infinitely large value, whereas when θ approaches $\dfrac{\pi}{2}$ from the side on which it is greater than $\dfrac{\pi}{2}$, $\tan \theta$ is negative and the absolute value of $\tan \theta$ can again be infinitely large.

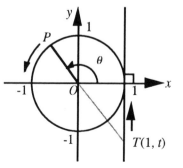

For $\theta > \pi$, and $\theta < 0$, the same variation is repeated.

When n is an integer, the value of $\tan \theta$ for

θ	0	...	$\frac{\pi}{2}$	...	π
Change in $\tan \theta$	0	↗ incr.	none	↗ incr.	0

$$\theta = \frac{\pi}{2} + n\pi$$

cannot be defined.

If we draw the graph of $\tan \theta$, we obtain the figure below.

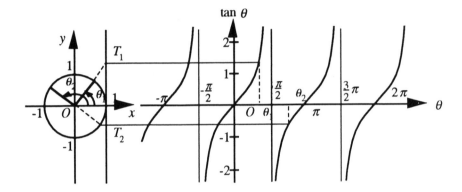

<u>Demonstration 1</u>   Use the graph of $y = \sin \theta$ to graph the following functions:

(1) $y = \sin (\theta + \frac{\pi}{6})$ (2) $y = 2 \sin \theta$

[**Solution**]   The graph of (1) is created by translating the graph of $y = \sin \theta$ by $-\frac{\pi}{6}$ along the horizontal axis.

The graph of (2) is created by expanding the graph of $y = \sin \theta$ along the vertical axis by a factor of 2.

Therefore, the graphs of (1) and (2) take the forms shown in the following figure:

Problem 1   Use the graph of $y = \sin \theta$ to graph the following functions:

(1) $y = -\sin \theta$ (2) $y = \sin 2\theta$

Problem 2   Use the graph of $y = \cos \theta$ to graph the following functions:

(1) $y = \cos \left(\theta - \dfrac{\pi}{4} \right)$ (2) $y = 2 \cos \dfrac{\theta}{2}$

Problem 3   Use the graph of $y = \tan \theta$ to graph the following functions:

(1) $y = -\tan \theta$ (2) $y = \dfrac{1}{2} \tan \theta$

Periodic Functions

When we examine the change in $\sin \theta$ and $\cos \theta$, the variation between 0 and 2π is repeated over intervals such as $\theta \le 0$ or $2\pi \le \theta$. This means that the change in these functions has a period of 2π.

In general, for a non-zero constant T, if

$$f(\theta + T) = f(\theta)$$

holds for all θ belonging to the domain, this function $f(\theta)$ is called a **periodic function**, and T is called the **period**.

For example, $\sin \theta$ and $\cos \theta$ are periodic functions with a period of 2π. Since the following equality holds by virtue of formula [VII'] on page 41,

$$\tan (\theta + \pi) = \tan \theta.$$

$\tan \theta$ is a periodic function with a period of π.

If T is the period of $f(x)$, then the graph of $y = f(x)$ is not changed by translating it T units to the right or left.

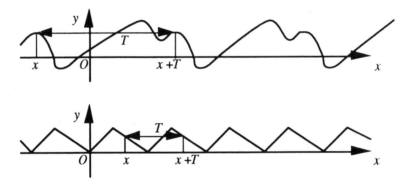

If T is the period of $f(x)$, then $2T$, $3T$, ..., $-T$, $-2T$, ... are also periods of $f(x)$. We refer to the minimum positive period as the **basic period**. The term "period" usually refers to the basic period.

Demonstration 2   Find the angle θ which satisfies $\tan \theta = 1$.

[Solution]   Since $\tan \theta$ has a period of π, we need to find α, the value of θ which satisfies $\tan \theta = 1$ on the interval

$$0 \leq \theta < \pi.$$

Angle α is formed by the positive ray of the x-axis and the straight line connecting the origin to the point on line $x = 1$ with a y-coordinate of 1, that is, the point $(1, 1)$. Therefore,

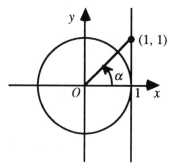

$$\alpha = \frac{\pi}{4}.$$

This fact, together with the periodic nature of $\tan \theta$, gives us the following expression for θ, when n is an arbitrary integer.

$$\theta = \alpha + n\pi = \frac{\pi}{4} + n\pi$$

Problem 4   Find the angle θ which satisfies the following equalites:

(1) $\tan \theta = -1$ 　　　　　　　　　　(2) $\tan \theta = \sqrt{3}$

Exercises

1.   We have an angle smaller than 360°. The terminal side of the angle created by multiplying the original angle by 7 is the same as the terminal side of the original angle. Find the original angle.

2.   Find the following values:

(1) $\sin (-210°)$ 　　　　　(2) $\cos 315°$ 　　　　　(3) $\tan 540°$

(4) $\cos \dfrac{13}{6} \pi$ 　　　　(5) $\tan (-\dfrac{19}{4} \pi)$ 　　　(6) $\sin (-\dfrac{20}{3} \pi)$

3.   Simplify the following expressions:

   (1) $\cos \theta + \sin \left(\dfrac{\pi}{2} - \theta \right) + \cos (\pi + \theta) + \sin \left(\dfrac{3}{2} \pi + \theta \right)$

   (2) $\dfrac{\cos \theta}{1 + \sin \theta} + \dfrac{\cos \theta}{1 - \sin \theta}$

4.   Find $\sin \theta$ and $\cos \theta$, if θ is an angle in the third quadrant and $\tan \theta = \dfrac{4}{3}$.

5.   Find $\sin \theta$ and $\tan (-\theta)$, if $\cos \theta = \dfrac{12}{13}$.

6.   Graph the following functions, and give their periods.

   (1) $y = -\cos \theta$ (2) $y = \sin 2\theta + 1$ (3) $y = \tan \left(\theta - \dfrac{\pi}{3} \right)$

7.   Find the value of x which satisfies the following equalities. We can assume that $0 \le x < 2\pi$.

   (1) $\sin x = -\dfrac{1}{2}$ (2) $\cos x = -1$ (3) $\sqrt{3} \tan x = -1$

8.   Find the range of values of x which satisfy the following inequalities. We can assume that $0 \le x < 2\pi$.

   (1) $\cos x > -\dfrac{1}{2}$ (2) $\tan x \le \sqrt{3}$

ADDITION THEOREMS

 Addition Theorems

When you work with trigonometric functions, the following formulas for the sine and cosine of the sum and difference of two angles are very important.

Sine and Cosine Addition Theorems

[I] $\sin(\alpha + \beta) = \sin \alpha \cos \beta + \cos \alpha \sin \beta$

   $\sin(\alpha - \beta) = \sin \alpha \cos \beta - \cos \alpha \sin \beta$

[II] $\cos(\alpha + \beta) = \cos \alpha \cos \beta - \sin \alpha \sin \beta$

   $\cos(\alpha - \beta) = \cos \alpha \cos \beta + \sin \alpha \sin \beta$

The formulas in [I] and [II] are called the sine and cosine **addition theorems**, respectively.

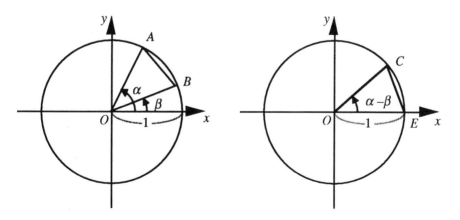

As you can see from the figures above, the distance between two points $A(\cos \alpha, \sin \alpha)$ and $B(\cos \beta, \sin \beta)$ is equal to the distance between two points $C(\cos(\alpha - \beta), \sin(\alpha - \beta))$ and $E(1, 0)$.

If we calculate AB^2 and CE^2 separately using the formula for the distance between two points, then we obtain

$$AB^2 = (\cos \alpha - \cos \beta)^2 + (\sin \alpha - \sin \beta)^2$$

$$= 2 - 2(\cos \alpha \cos \beta + \sin \alpha \sin \beta);$$

$$CE^2 = \{\cos (\alpha - \beta) - 1\}^2 + \{\sin (\alpha - \beta) - 0\}^2$$

$$= 2 - 2 \cos (\alpha - \beta).$$

Since $AB^2 = CE^2$,

$$\cos (\alpha - \beta) = \cos \alpha \cos \beta + \sin \alpha \sin \beta.$$

We have derived the second formula in [II]. If we substitute $-\beta$ for β, we obtain

$$\cos (\alpha + \beta) = \cos \alpha \cos (-\beta) + \sin \alpha \sin (-\beta)$$

$$= \cos \alpha \cos \beta - \sin \alpha \sin \beta.$$

If we substitute $\frac{\pi}{2} - \alpha$ for α, we get

$$\cos \left(\frac{\pi}{2} - (\alpha - \beta) \right) = \cos\left(\frac{\pi}{2} - \alpha \right) \cos \beta - \sin \left(\frac{\pi}{2} - \alpha \right) \sin \beta.$$

Therefore,

$$\sin (\alpha - \beta) = \sin \alpha \cos \beta - \cos \alpha \sin \beta.$$

Furthermore, if we substitute $-\beta$ for β, we obtain

$$\sin (\alpha + \beta) = \sin \alpha \cos (-\beta) - \cos \alpha \sin (-\beta)$$

$$= \sin \alpha \cos \beta + \cos \alpha \sin \beta.$$

Now we have derived all four of the sine and cosine addition theorems.

Example $\sin 75° = \sin (45° + 30°) = \sin 45° \cos 30° + \cos 45° \sin 30°$

$$= \frac{1}{\sqrt{2}} \cdot \frac{\sqrt{3}}{2} + \frac{1}{\sqrt{2}} \cdot \frac{1}{2} = \frac{\sqrt{6} + \sqrt{2}}{4}$$

$\sin 15° = \sin (45° - 30°) = \sin 45° \cos 30° - \cos 45° \sin 30°$

$$= \frac{1}{\sqrt{2}} \cdot \frac{\sqrt{3}}{2} - \frac{1}{\sqrt{2}} \cdot \frac{1}{2} = \frac{\sqrt{6} - \sqrt{2}}{4}$$

Problem 1   Find $\cos 75°$ and $\cos 15°$ as in the above Example.

Problem 2   Find $\sin 105°$ and $\cos 105°$, using $105° = 60° + 45°$.

Demonstration   Find the value of $\sin (\alpha + \beta)$, if $\sin \alpha = \frac{4}{5}$

and $\cos \beta = -\frac{8}{17}$. We can assume that $0 < \alpha < \frac{\pi}{2}$

and $\frac{\pi}{2} < \beta < \pi$.

[Solution] $\cos^2 \alpha = 1 - \sin^2 \alpha = 1 - (\frac{4}{5})^2 = (\frac{3}{5})^2$

Since $0 < \alpha < \frac{\pi}{2}$, $\cos \alpha > 0$. Therefore, $\cos \alpha = \frac{3}{5}$.

Next,

$$\sin^2 \beta = 1 - \cos^2 \beta = 1 - (-\frac{8}{17})^2 = (\frac{15}{17})^2.$$

Since $\frac{\pi}{2} < \beta < \pi$, $\sin \beta > 0$. Therefore, $\sin \beta = \frac{15}{17}$.

Thus,

$$\sin (\alpha + \beta) = \sin \alpha \cos \beta + \cos \alpha \sin \beta$$

$$= \frac{4}{5} \cdot (-\frac{8}{17}) + \frac{3}{5} \cdot \frac{15}{17} = \frac{13}{85}.$$

Problem 3   Find the values of $\sin(\alpha - \beta)$, $\cos(\alpha + \beta)$, and $\cos(\alpha - \beta)$ for α and β in the above Demonstration.

The tangent addition theorem given below can be derived from the sine and cosine addition theorems.

Tangent Addition Theorem

[III] $\tan(\alpha + \beta) = \dfrac{\tan \alpha + \tan \beta}{1 - \tan \alpha \tan \beta}$

$\tan(\alpha - \beta) = \dfrac{\tan \alpha - \tan \beta}{1 + \tan \alpha \tan \beta}$

Problem 4   Derive the first formula in [III] from theorems [I] and [II].

Problem 5   Find the tangents of 15°, 75°, and 105° using the values of the trigonometric functions of 30°, 45°, and 60°.

Application of the Addition Theorems

The following formulas for double angles can be derived from addition theorems [I], [II], and [III].

Double Angle Theorem

[IV] $\sin 2\alpha = 2 \sin \alpha \cos \alpha$

$\cos 2\alpha = \cos^2 \alpha - \sin^2 \alpha$

$= 1 - 2 \sin^2 \alpha = 2 \cos^2 \alpha - 1$

$\tan 2\alpha = \dfrac{2 \tan \alpha}{1 - \tan^2 \alpha}$

The first formula in [IV] can be derived in the following way:

$$\sin 2\alpha = \sin (\alpha + \alpha)$$
$$= \sin \alpha \cos \alpha + \cos \alpha \sin \alpha$$
$$= 2 \sin \alpha \cos \alpha.$$

Problem 1 Prove the formulas for the cosine and tangent of double angles.

Problem 2 Prove the following equalities:

(1) $(\sin \alpha + \cos \alpha)^2 = 1 + \sin 2\alpha$

(2) $\cos^4 \alpha - \sin^4 \alpha = \cos 2\alpha$

(**Demonstration 1**)   Find $\sin 2\alpha$ and $\cos 2\alpha$, if $\cos \alpha = -\dfrac{4}{5}$. We can

assume that $\dfrac{\pi}{2} < \alpha < \pi$.

[**Solution**]   Since $\dfrac{\pi}{2} < \alpha < \pi$, $\sin \alpha > 0$.

$$\sin \alpha = \sqrt{1 - \cos^2 \alpha} = \sqrt{1 - \left(-\dfrac{4}{5}\right)^2} = \dfrac{3}{5}$$

Therefore,

$$\sin 2\alpha = 2 \sin \alpha \cos \alpha = 2 \cdot \dfrac{3}{5} \cdot \left(-\dfrac{4}{5}\right) = -\dfrac{24}{25},$$

$$\cos 2\alpha = 2 \cos^2 \alpha - 1 = 2 \cdot \left(-\dfrac{4}{5}\right)^2 - 1 = \dfrac{7}{25}.$$

(**Problem 3**)   Find $\sin 2\alpha$, $\cos 2\alpha$, and $\tan 2\alpha$, if $\sin \alpha = -\dfrac{1}{2}$. We can assume

that $-\dfrac{\pi}{2} < \alpha < 0$.

(**Problem 4**)   Prove the following equalities:

(1) $\sin 3\alpha = 3 \sin \alpha - 4 \sin^3 \alpha$

(2) $\cos 3\alpha = 4 \cos^3 \alpha - 3 \cos \alpha$

(**Problem 5**)   Derive the following equalities from the formula for the cosine of a double angle:

$$\sin^2 \dfrac{\alpha}{2} = \dfrac{1 - \cos \alpha}{2},$$

$$\cos^2 \dfrac{\alpha}{2} = \dfrac{1 + \cos \alpha}{2},$$

$$\tan^2 \dfrac{\alpha}{2} = \dfrac{1 - \cos \alpha}{1 + \cos \alpha}.$$

Demonstration 2   The coordinates of point P are $(3, 2)$. We take α as the angle formed by terminal side OP and the positive ray of the x-axis. Prove the following equality:

$$3 \sin \theta + 2 \cos \theta = \sqrt{13} \sin (\theta + \alpha).$$

[Proof]   Since $P(3, 2)$,

$$OP = \sqrt{3^2 + 2^2} = \sqrt{13}$$

Therefore,

$$\sin \alpha = \frac{2}{\sqrt{13}}, \quad \cos \alpha = \frac{3}{\sqrt{13}}.$$

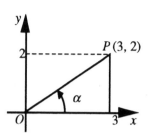

Using these results, we obtain

$$3 \sin \theta + 2 \cos \theta = \sqrt{13} \left(\frac{3}{\sqrt{13}} \sin \theta + \frac{2}{\sqrt{13}} \cos \theta \right)$$

$$= \sqrt{13} (\cos \alpha \sin \theta + \sin \alpha \cos \theta)$$

$$= \sqrt{13} \sin (\theta + \alpha).$$

Problem 6   Find r and α to satisfy

$$\sin \theta + \cos \theta = r \sin (\theta + \alpha)$$

by referring to the figure to the right.

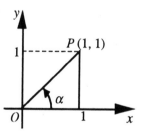

Problem 7   Show that if we pick α as in the figure to the right, the following equality holds:

$$a \sin \theta + b \cos \theta = \sqrt{a^2 + b^2} \sin (\theta + \alpha).$$

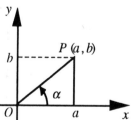

Problem 8   Convert the following expressions into the form $r \sin (\theta + \alpha)$:

(1) $\sin \theta + 2 \cos \theta$

(2) $\sqrt{3} \sin \theta - \cos \theta$

Demonstration 3   Find the maximum and minimum values of

$y = \sin \theta + \sqrt{3} \cos \theta$. Then graph the function.

[**Solution**]   Using the figure to the right,

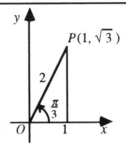

$$y = \sin \theta + \sqrt{3} \cos \theta$$

$$= 2 \sin \left(\theta + \frac{\pi}{3} \right).$$

Since the maximum value of $\sin \left(\theta + \frac{\pi}{3} \right)$ is 1 and the minimum value is -1, this function takes on a maximum value of 2 and a minimum value of -2.

The graph of this function is created by translating the graph of $y = \sin \theta$ by $-\frac{\pi}{3}$ along the horizontal axis, and then expanding it by a factor of two.

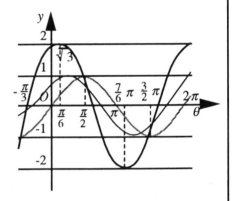

Problem 9   Graph the following functions:

(1) $y = \sin \theta + \cos \theta$ (2) $y = \sin \theta - \cos \theta$

Demonstration 4   Prove that the following equalities hold:

$$\sin A + \sin B \;=\; 2\sin\frac{A+B}{2}\,\cos\frac{A-B}{2}\,,$$

$$\sin A - \sin B \;=\; 2\cos\frac{A+B}{2}\,\sin\frac{A-B}{2}\,.$$

[Proof] $\sin(\alpha+\beta) = \sin\alpha\cos\beta + \cos\alpha\sin\beta$

$\sin(\alpha-\beta) = \sin\alpha\cos\beta - \cos\alpha\sin\beta$

Hence,

$$\sin(\alpha+\beta) + \sin(\alpha-\beta) \;=\; 2\sin\alpha\cos\beta,$$

$$\sin(\alpha+\beta) - \sin(\alpha-\beta) \;=\; 2\cos\alpha\sin\beta.$$

If we set $\alpha+\beta=A$ and $\alpha-\beta=B$, then we obtain

$$\alpha = \frac{A+B}{2}\,,\quad \beta = \frac{A-B}{2}\,.$$

Therefore,

$$\sin A + \sin B \;=\; 2\sin\frac{A+B}{2}\,\cos\frac{A-B}{2}\,,$$

$$\sin A - \sin B \;=\; 2\cos\frac{A+B}{2}\,\sin\frac{A-B}{2}\,.$$

Problem 10   Prove the following equalities, as in Demonstration 4.

$$\cos A + \cos B \;=\; 2\cos\frac{A+B}{2}\,\cos\frac{A-B}{2}$$

$$\cos A - \cos B \;=\; -2\sin\frac{A+B}{2}\,\sin\frac{A-B}{2}$$

Exercises

1. Find the value of $\sin(\alpha + \beta)$ and $\cos(\alpha - \beta)$, if $\sin \alpha = \dfrac{3}{5}$ and $\cos \beta = -\dfrac{5}{13}$. We can assume that $0 < \alpha < \dfrac{\pi}{2}$ and $\dfrac{\pi}{2} < \beta < \pi$.

2. Find the values of $\sin 2\alpha$, $\cos 2\alpha$, and $\tan 2\alpha$ for $\sin \alpha = -0.8$. We can assume that $-\dfrac{\pi}{2} < \alpha < 0$.

3. Prove the following equalities:

    (1) $\sin(\alpha + \beta)\sin(\alpha - \beta) = \sin^2 \alpha - \sin^2 \beta = \cos^2 \beta - \cos^2 \alpha$

    (2) $\tan\left(\dfrac{\pi}{4} + \theta\right) = \dfrac{1 + \tan\theta}{1 - \tan\theta}$

    (3) $\dfrac{\sin 2\theta}{1 + \cos 2\theta} = \tan\theta$

4. Find the maximum and minimum values of the following functions:

    (1) $y = 5\sin x + 3\cos x$ (2) $y = 2\sin x - \cos x$

5. Point P moves along the y-axis. The coordinate y of point P after t seconds can be expressed as $y = 3\sin\dfrac{\pi}{6}t$. Graph this function on the interval $0 \le t \le 12$.

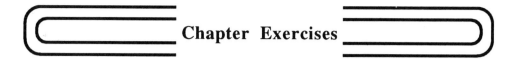

Chapter Exercises

A

1. Find the values of the following trigonometric functions:

    (1) $\cos \dfrac{7}{6}\pi$

    (2) $\tan \dfrac{25}{4}\pi$

    (3) $\sin\left(-\dfrac{5}{3}\pi\right)$

2. In what quadrant can we find the angles which satisfy simultaneously the following pairs of inequalities?

    (1) $\begin{cases} \sin\theta < 0 \\ \cos\theta < 0 \end{cases}$

    (2) $\begin{cases} \cos\theta > 0 \\ \tan\theta < 0 \end{cases}$

3. Express the following values in terms of k, where $\sin\alpha = k$, if α is an angle in the second quadrant.

    (1) $\sin(3\pi + \alpha)$

    (2) $\cos(-\alpha)$

    (3) $\cos(\pi - \alpha)$

    (4) $\tan\left(\dfrac{\pi}{2} + \alpha\right)$

4. Find the values of the following expressions:

    (1) $\sqrt{2}\cos\left(-\dfrac{5}{4}\pi\right) + 3\sin\dfrac{3}{2}\pi - \sin\left(-\dfrac{\pi}{6}\right)$

    (2) $\sin^{2}\dfrac{2}{3}\pi + \sin\left(-\dfrac{7}{3}\pi\right)\cos\left(-\dfrac{5}{6}\pi\right)$

5. Find the following values, if α is an angle in the second quadrant, β lies in the fourth quadrant, $\sin\alpha = 0.2$, and $\cos\beta = 0.6$:

    (1) $\cos(\alpha - \beta)$

    (2) $\sin(\alpha + 2\beta)$

6. Find the values of θ which satisfy the following equalities. We can assume that $0 \le \theta < 2\pi$.

    (1) $2\sin\theta = -\sqrt{3}$

    (2) $\cos 2\theta = -1$

7.   Graph the following functions:

(1) $y = \cos (x + \frac{\pi}{2})$

(2) $y = 2 \sin (2x + \frac{\pi}{3})$

$\mathbb{B}$

1.   Prove that the length of arc l and the area S of the sector in the figure to the right can be expressed as

$$l = r\theta, \qquad S = \frac{1}{2}r^2\theta$$

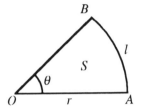

where θ is the central angle and r is the radius. The angle θ is given in radian measure.

2.   Prove the following equalities:

(1) $\dfrac{1 + \cos \theta}{1 - \sin \theta} - \dfrac{1 - \cos \theta}{1 + \sin \theta} = \dfrac{2(1 + \tan \theta)}{\cos \theta}$

(2) $\dfrac{\sin \theta}{1 + \cos \theta} + \dfrac{1 + \cos \theta}{\sin \theta} = \dfrac{2}{\sin \theta}$

3.   Find the value of $\sin \theta$ and $\cos \theta$ for $\tan \theta = 2 - \sqrt{3}$.

4.   Find the period of the following functions:

(1) $y = \dfrac{1}{3} \tan \dfrac{x}{3}$

(2) $y = \sin 2x \cos 2x$

5.   Convert the following functions into the form $r \sin (x + \theta)$, and find the maximum and minimum values.

(1) $\sin (x + \dfrac{\pi}{4}) + \cos (x - \dfrac{\pi}{4})$

(2) $\sin x + 2 \cos (x - \dfrac{\pi}{6})$

6.   Graph $y = |\sin x|$ based on the graph of $y = \sin x$.

7.   Find the value of $\sin \theta$, if $\sin \theta = 2 \sin^2 \theta$. Then find the value of θ. We can assume that $0 \le \theta < 2\pi$.

8.   (1)   Find the values of

$$\sin 75° + \sin 15°, \quad \sin 75° - \sin 15°$$

using the formulas given in Demonstration 4 on page 59.

(2)   Find the values of $\sin 15°$ and $\sin 75°$ using the results of (1).

CHAPTER 3

PROGRESSIONS

SECTION 1.   PROGRESSIONS
SECTION 2.   MATHEMATICAL INDUCTION AND PROGRESSIONS

Mathematics is said to be "the study of infinity."

In studying progressions you will encounter the concept of infinity for the first time since elementary school. Infinite progressions will not be the major topic of this chapter, but the idea of "any finite number," which could be described as one step before infinity, will be considered.

Progressions are functions which relate real numbers or complex numbers to each natural number. A function which relates one point to each natural number is called a sequence of points.

If we regard a progression as a kind of function, an arithmetic progression is a linear function, and a geometric progression is an exponential function.

The study of progressions whose domain is the set of all natural numbers – the simplest infinite set – is the basis for the study of more general functions.

① PROGRESSIONS

Progressions

If we write down the natural numbers one at a time, starting with 1, we get

$$1, 2, 3, 4, 5, \dots .$$

If we write down the numbers obtained by repeatedly multiplying 3 by the preceding number, starting with 1, we obtain

$$1, 3, 9, 27, 81, \dots .$$

A sequence such as these, in which numbers are arranged according to a certain rule, is called a **progression**.

Each number in a progression is called a **term**. Starting from the beginning, the numbers are referred to as the **first term**, **second term**, **third term**, and so on.

In general, the term in the nth place from the beginning is called the **nth term**.

A progression in which the first term is a_1, the second term is a_2, the third term is a_3, ..., the nth term is a_n, can be expressed as

$$a_1, a_2, a_3, \dots, a_n, \dots .$$

The usual mathematical notation for this progression is $\{a_n\}$.

Example 1 In a progression in which the nth term a_n is $4n - 3$,

the first term is $\quad a_1 = 4 \times 1 - 3 = 1,$

the second term is $\quad a_2 = 4 \times 2 - 3 = 5,$

the third term is $\quad a_3 = 4 \times 3 - 3 = 9,$

$\dots .$

Example 2 If $a_n = \dfrac{1}{2^{n-1}}$, then progression $\{a_n\}$ is

$$1, \frac{1}{2}, \frac{1}{4}, \frac{1}{8}, \frac{1}{16}, \cdots .$$

Expressions in n, such as $4n - 3$ and $\dfrac{1}{2^{n-1}}$ in Examples 1 and 2, express the nth term of each progression.

In these expressions, if we take $n = 1$, we obtain the first term, $n = 2$ gives us the second term, $n = 3$ gives us the third term, and so on.

Such an expression in n, representing the nth term, is called the **general term** of the progression.

Problem Write the first five terms of the following progressions:

(1) $\{1 - 3n\}$ (2) $\{n^2 - n + 1\}$ (3) $\{(-1)^n\}$

A progression with a finite number of terms is called a **finite progression**. In a finite progression, we often make special reference to the **number of terms** and the **last term**.

For example,

$$3, 6, 9, 12, 15, 18, 21, 24, 27, 30$$

is a finite progression; the number of terms is 10 and the last term is 30.

 Arithmetic Progressions

The progression

$$3, 8, 13, 18, 23, \ldots$$

is created by adding 5 to the preceding number, starting with 3 as the first term.

A progression created by adding a fixed number to the preceding number over and over, starting from the first term, is called an **arithmetic progression**, and the fixed number is called the **common difference**.

The following relations hold between the nth term a_n and the $(n+1)$th term a_{n+1} of an arithmetic progression with common difference d.

$$a_{n+1} = a_n + d \quad \text{or} \quad a_{n+1} - a_n = d$$

(Example)   The progression

$$3, \ 7, \ 11, \ 15, \ 19, \ 23, \ ...$$

is an arithmetic progression in which the first term is 3 and the common difference is 4.

A progression with a first term of 14 and a common difference of -6 is

$$14, \ 8, \ 2, \ -4, \ -10, \ -16, \ ... \ .$$

The first several terms of a progression in which the first term is a and the common difference is d are

$$a_1 = a, \ a_2 = a + d, \ a_3 = a + 2d, \ a_4 = a + 3d, \ ... \ .$$

The General Term of an Arithmetic Progression

The nth term a_n of an arithmetic progression with a first term of a and a common difference of d is

$$a_n = a + (n - 1)d.$$

(Problem 1)   (1)   Find the 20th term of an arithmetic progression in which the first term is 4 and the common difference is 3.

(2)   Find the 25th term of an arithmetic progression in which the first term is 7 and the common difference is $-\dfrac{1}{3}$.

(Problem 2)   Find the general term of the following arithmetic progressions:

(1)   23, 30, 37, 44, ...   (2)   2, $\dfrac{5}{4}$, $\dfrac{1}{2}$, $-\dfrac{1}{4}$, ...

Problem 3   Which term of an arithmetic progression with a first term of 8 and a common difference of -3 will be -37?

Demonstration 1   Find the first term and the common difference of an arithmetic progression in which the 4th term is 14 and the 9th term is 54.

[Solution]   Take a as first term of this progression and d as the common difference, and then

$$a_4 = a + 3d = 14, \qquad a_9 = a + 8d = 54.$$

Solving these equalities, we obtain

$$a = -10, \qquad d = 8.$$

Therefore, the first term is -10 and the common difference is 8.

Problem 4   Find the general term of an arithmetic progression in which the third term is -4 and the 10th term is 38.

Problem 5   Find the common difference of an arithmetic progression in which the first term is 3, the last term is 94, and the number of terms is 15. Then find the 5th term and the 10th term.

Demonstration 2   Prove that a progression in which the nth term is expressed as $5n + 3$ is an arithmetic progression.

[Proof]   Since

$$a_n = 5n + 3$$

and

$$a_{n+1} = 5(n + 1) + 3 = 5n + 8,$$

it follows that

$$a_{n+1} - a_n = 5.$$

Thus, this progression is an arithmetic progression with a common difference of 5.

Problem 6   Prove that the progression $\{pn + q\}$ is an arithmetic progression, provided that p and q are constants.

Problem 7   Prove that if two progressions $\{a_n\}$ and $\{b_n\}$ are arithmetic progressions, then the progression $\{a_n + b_n\}$ is also an arithmetic progression.

Next, let's consider the first n terms of an arithmetic progression.

Take S_n as the sum of the first n terms of an arithmetic progression in which the first term is a and the common difference is d.

$$S_n = a + (a + d) + (a + 2d) + \ldots + \{a + (n - 1)d\} \tag{1}$$

If we take l as the last term of this progression and rewrite the right side by reversing the order of the terms, then

$$S_n = l + (l - d) + (l - 2d) + \ldots + \{l - (n - 1)d\}. \tag{2}$$

Adding up the corresponding sides of (1) and (2), we obtain

$$2S_n = (a + l) + (a + l) + (a + l) + \ldots + (a + l).$$

The right side of this expression is equivalent to n times $(a + l)$. Therefore,

$$2S_n = n(a + l).$$

Thus,

$$S_n = \frac{n(a + l)}{2}.$$

Substituting $l = a + (n - 1)d$ into this expression, we obtain

$$S_n = \frac{n\{2a + (n - 1)d\}}{2}.$$

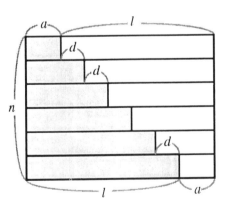

The Sum of an Arithmetic Progression

The sum S_n of an arithmetic progression with a first term of a, a last term of l, and a total of n terms is

$$S_n = \frac{n(a + l)}{2}.$$

The sum S_n of the first n terms of an arithmetic progression with a first term of a and a common difference of d is

$$S_n = \frac{n\{2a + (n - 1)d\}}{2}.$$

Problem 8 Find the sum of each of the following arithmetic progressions:

(1) The first term is 7, the last term is 61, and there are 10 terms.

(2) The first term is -10, the common difference is 4, and there are 13 terms.

(3) The first term is 21, the common difference is -6, and the last term is -117.

Problem 9 Prove that the following equalities hold:

(1) $1 + 2 + 3 + ... + n = \dfrac{n(n + 1)}{2}$

(2) $1 + 3 + 5 + ... + (2n - 1) = n^2$

Problem 10 (1) Find the sum of the natural numbers from 1 through 100.

(2) Find the sum of all the multiples of 3 among the natural numbers through 200.

Problem 11 If the sum of the first n terms of an arithmetic progression is 297, the first term is 45, and the common difference is -3, what is the value of n?

Geometric Progressions

A progression created by multiplying the preceding term by a fixed number again and again, starting from the first term, is called a **geometric progression**, and the fixed number is called the **common ratio**.

Therefore, the following relation holds between the nth term a_n and the $(n+1)$th term a_{n+1} of a geometric progression with a common ratio of r:

$$a_{n+1} = a_n r.$$

As a special case, when no term is equal to 0, we have

$$\frac{a_{n+1}}{a_n} = r.$$

Example   The progression 1, 2, 4, 8, 16, ... is a geometric progression in which the first term is 1 and the common ratio is 2. A geometric progression with a first term of 27 and a common ratio of $-\dfrac{1}{3}$ is

$$27, -9, 3, -1, \frac{1}{3}, -\frac{1}{9}, \ \dots \ .$$

The first several terms of a geometric progression with a first term of a and a common ratio of r are

$$a_1 = a, \ a_2 = ar, \ a_3 = ar^2, \ a_4 = ar^3, \ \dots \ .$$

The General Term of a Geometric Progression

The nth term a_n of a geometric progression with a first term of a and a common ratio of r is

$$a_n = ar^{n-1}.$$

Problem 1   (1) Find the 6th term of a geometric progression in which the first term is 1 and the common ratio is 2.

(2) Find the 5th term of a geometric progression in which the first term is 3 and the common ratio is $-\dfrac{1}{3}$.

Problem 2   State the common ratio of the following geometric progressions. Then find the general term.

(1) $\sqrt{2}, 2, 2\sqrt{2}, 4, \dots$ (2)   1, -3, 9, -27, 81, ...

(3)   1, -1, 1, -1, 1, ...

Problem 3   State the first term and the common ratio of a geometric progression in which the general term is $\dfrac{3^{n+1}}{2^n}$.

Problem 4   Find the first term and the common ratio of a geometric progression in which the third term is 4 and the 5th term is 36.

Next, let's consider the sum of the first n terms of a geometric progression.

If we take S_n as the sum of the first n terms of a geometric progression in which the first term is a and the common ratio is r, then

$$S_n = a + ar + ar^2 + \dots + ar^{n-2} + ar^{n-1}.$$ (1)

Multiplying both sides of this equation by r, we obtain

$$rS_n = ar + ar^2 + ar^3 + \dots + ar^{n-1} + ar^n.$$ (2)

Subtracting the left side and the right side of (2) from the corresponding sides of (1), we get

$$(1 - r)S_n = a - ar^n.$$

If $r \ne 1$, then

$$S_n = \frac{a(1 - r^n)}{1 - r} .$$

If $r = 1$, then from (1), we obtain

$$S_n = a + a + \dots a = na.$$

The Sum of a Geometric Progression

The sum of the first n terms of a geometric progression with a first term of a and a common ratio of r is

$$\text{for } r \neq 1, \quad S_n = \frac{a(1 - r^n)}{1 - r} .$$

$$\text{for } r = 1, \quad S_n = na.$$

Problem 5   Find the sum of each of the following progressions:

(1)   The first term is 3, the common ratio is 2, and there are 7 terms.

(2)   The first term is 1, the common ratio is -3, and there are 6 terms.

(3)   The first term is 5, the common ratio is $\frac{1}{2}$, and there are 5 terms.

Problem 6   Find the sum of the first n terms of the following geometric progressions:

(1)   13, 52, 208, 832, ...   (2) $\sqrt{3}$, -1, $\frac{1}{\sqrt{3}}$, $-\frac{1}{3}$, ...

(**Demonstration 1**) If a fixed amount of a yen is added to savings at the beginning of each quarter, find the cumulative balance at the end of the nth quarter. Assume that the interest rate for a quarter is r, and that the interest is compounded quarterly.

[Solution] At the end of the nth quarter the amount deposited in the savings account for the first quarter, second quarter, third quarter, ..., nth quarter can be expressed in the following form:

$$a(1 + r)^n \text{ yen, } a(1 + r)^{n-1} \text{ yen, } a(1 + r)^{n-2} \text{ yen,}$$
$$..., a(1 + r) \text{ yen.}$$

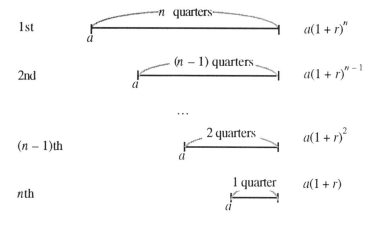

If we take S yen as the sum of these amounts, then

$$S = a(1 + r) + a(1 + r)^2 + ... + a(1 + r)^n.$$

This is the sum of a geometric progression in which the first term is $a(1 + r)$, the common ratio is $(1 + r)$, and the number of terms is n. The common ratio is not equal to 1, and therefore,

$$S = \frac{a(1 + r)\{1 - (1 + r^n)\}}{1 - (1 + r)} = \frac{a(1 + r)\{(1 + r)^n - 1\}}{r}.$$

(**Problem 7**) If we borrow A yen at the beginning of a certain year, then we must pay back a yen at the end of each year for the next n years in order to pay off the loan. Assuming an annual compound interest rate of r, express a in terms of $A, r,$ and n.

(**Demonstration 2**)   Find the following sum, provided that $x \neq 1$.

$$1 + 2x + 3x^2 + \dots + nx^{n-1}$$

[**Solution**]   Set

$$S_n = 1 + 2x + 3x^2 + \dots + (n-1)x^{n-2} + nx^{x-1}. \tag{1}$$

Multiplying both sides by x, we obtain

$$xS_n = x + 2x^2 + 3x^3 + \dots + (n-1)x^{n-1} + nx^n. \tag{2}$$

Subtracting (2) from (1), we get

$$(1-x)S_n = 1 + x + x^2 + \dots + x^{n-1} - nx^n.$$

Since $x \neq 1$, we can rearrange the right side to get

$$(1-x)S_n = \frac{1-x^n}{1-x} - nx^n = \frac{1-(n-1)x^n + nx^{n+1}}{1-x}.$$

Thus,

$$S_n = \frac{1-(n+1)x^n + nx^{n+1}}{(1-x)^2}.$$

(**Problem 8**)   Find the sum of the first n terms of the following progression:

$$1 \cdot 2, \ 4 \cdot 2^2, \ 7 \cdot 2^3, \ 10 \cdot 2^4, \ 13 \cdot 2^5, \ \dots .$$

Various Types of Progressions

There are progressions besides arithmetic and geometric progressions in which the general term or the sum of the first n terms can easily be found. Let's consider some of these other progressions.

Demonstration 1   Find the following sum:

$$1^2 + 2^2 + 3^2 + \ldots + n^2.$$

[Solution]   In the identity $(k + 1)^3 - k^3 = 3k^2 + 3k + 1$,

if $k = 1$, then $2^3 - 1^3 = 3 \cdot 1^2 + 3 \cdot 1 + 1$;

if $k = 2$, then $3^3 - 2^3 = 3 \cdot 2^2 + 3 \cdot 2 + 1$;

...

if $k = n$, then $(n + 1)^3 - n^3 = 3 \cdot n^2 + 3 \cdot n + 1$.

Adding the corresponding sides of these equalities, we obtain

$(n + 1)^3 - 1^3$

$= 3(1^2 + 2^2 + \ldots + n^2) + 3(1 + 2 + \ldots + n) + n$

$= 3(1^2 + 2^2 + \ldots + n^2) + \dfrac{3n(n + 1)}{2} + n.$

Therefore,

$$3(1^2 + 2^2 + \ldots + n^2) = (n + 1)^3 - \dfrac{3n(n + 1)}{2} - (n + 1)$$

$$= \dfrac{n(n + 1)(2n + 1)}{2}.$$

Thus,

$$1^2 + 2^2 + \ldots + n^2 = \dfrac{n(n + 1)(2n + 1)}{6}.$$

(Problem 1) Find the following values using the result of Demonstration 1:

(1) $1^2 + 2^2 + 3^2 + ... + 10^2$

(2) $8^2 + 9^2 + 10^2 + ... + 15^2$

(Problem 2) Prove the following equality, using the identity
$(k + 1)^4 - k^4 = 4k^3 + 6k^2 + 4k + 1$ as in Demonstration 1.

$$1^3 + 2^3 + 3^3 + ... + n^3 = (\frac{n(n + 1)}{2})^2$$

Use this equality to find the following sum:

$$6^3 + 7^3 + 8^3 + ... + 13^3.$$

The Symbol $\sum$ for the Sum of a Progression

We can write the sum of the first n terms of a progression $a_1, a_2, a_3, ..., a_n ...$ using the symbol $\sum$* as

$$\sum_{k=1}^{n} a_k.$$

Thus,

$$\sum_{k=1}^{n} a_k = a_1 + a_2 + a_3 + ... + a_n.$$

Using this symbol, $1^2 + 2^2 + 3^2 + ... + n^2$ can be represented as $\sum_{k=1}^{n} k^2$. This expression can also be written using letters other than k such as $\sum_{i=1}^{n} i^2$ or $\sum_{j=1}^{n} j^2$.

---

*$\sum$ is the Greek letter corresponding to S, the first letter of *Sum*, and is read "sigma" or, if you prefer, as "sum".

Moreover, $\displaystyle\sum_{k=3}^{6}(2k+3)$ expresses the sum $9 + 11 + 13 + 15$ given by substituting 3, 4, 5, and 6 for k in $2k + 3$.

Furthermore, $\displaystyle\sum_{k=1}^{n}1$ is the sum of the first n terms of the progression 1, 1, 1, ..., 1, ...

Therefore,

$$\sum_{k=1}^{n}1 = \underbrace{1 + 1 + 1 + ... + 1}_{n \text{ times}} = n \ .$$

Analogously, if c is a constant, then

$$\sum_{k=1}^{n}c = cn.$$

Problem 3   Express the following sums using $\displaystyle\sum$.

(1) $1^3 + 2^3 + 3^3 + ... + n^3$

(2) $5 + 9 + 13 + 17 + ... + 41$

Problem 4   Rewrite the following sums as the sum of all the terms without using $\displaystyle\sum$.

(1) $\displaystyle\sum_{k=1}^{8}(3k+1)$ (2) $\displaystyle\sum_{i=0}^{4}(\frac{1}{2})^i$ (3) $\displaystyle\sum_{j=1}^{n}j(j+1)$

We can use $\sum$ to formulate the following expressions for the sum of the given progressions:

The Sum of a Power

$$\sum_{k=1}^{n} k = \frac{n(n+1)}{2} \qquad \sum_{k=1}^{n} k^2 = \frac{n(n+1)(2n+1)}{6}$$

$$\sum_{k=1}^{n} k^3 = \{\frac{n(n+1)}{2}\}^2$$

Given the two progressions

$$a_1, a_2, a_3, ..., a_n$$

$$b_1, b_2, b_3, ..., b_n$$

the following generalization holds:

$$(a_1 + b_1) + (a_2 + b_2) + ... + (a_n + b_n)$$

$$= (a_1 + a_2 + ... + a_n) + (b_1 + b_2 + ... + b_n).$$

Also, for a constant c,

$$ca_1 + ca_2 + ... + ca_n = c(a_1 + a_2 + ... + a_n).$$

If we express these formulas using $\sum$, we obtain the following formulas.

The Properties of $\sum$

$$\sum_{k=1}^{n} (a_k + b_k) = \sum_{k=1}^{n} a_k + \sum_{k=1}^{n} b_k$$

$$\sum_{k=1}^{n} ca_k = c \sum_{k=1}^{n} a_k$$

Demonstration 2 Find the sum of the first n terms of a progression in which the nth term is expressed as $n^2 + 3n - 4$.

[Solution] Since the kth term of this progression is $k^2 + 3k - 4$, take S_n as the sum we want to find, and then

$$S_n = \sum_{k=1}^{n} (k^2 + 3k - 4) = \sum_{k=1}^{n} k^2 + \sum_{k=1}^{n} 3k + \sum_{k=1}^{n} (-4)$$

$$= \sum_{k=1}^{n} k^2 + 3\sum_{k=1}^{n} k - 4\sum_{k=1}^{n} 1$$

$$= \frac{n(n+1)(2n+1)}{6} + \frac{3n(n+1)}{2} - 4n$$

$$= \frac{1}{3} n(n-1)(n+7).$$

Problem 5 Find the following sums:

(1) $\displaystyle\sum_{k=1}^{n} (5k + 1)$

(2) $\displaystyle\sum_{i=1}^{n-1} (i+1)(i-2)$

Problem 6 Find the following sums:

(1) $1 \cdot 2 + 2 \cdot 3 + 3 \cdot 4 + \ldots + n(n+1)$

(2) $1 \cdot 2 \cdot 3 + 2 \cdot 3 \cdot 4 + \ldots + n(n+1)(n+2)$

Problem 7 Find the sum of the first n terms of the following progressions:

(1) $1 \cdot 3, \ 2 \cdot 4, \ 3 \cdot 5, \ 4 \cdot 6, \ 5 \cdot 7, \ \ldots$

(2) $1^2 \cdot 2, \ 2^2 \cdot 5, \ 3^2 \cdot 8, \ 4^2 \cdot 11, \ 5^2 \cdot 14, \ \ldots$

We are given a progression $\{a_n\}$. The progression $\{b_n\}$ created by

$$b_n = a_{n+1} - a_n \quad (n = 1, 2, 3, ...).$$

is called the **progression of differences** of the original progression.

$b_1 = a_2 - a_1$

$b_2 = a_3 - a_2$

$\quad a_1 \quad a_2 \quad a_3 \quad a_4 \quad \cdots \quad a_{n-1} \quad a_n$

$\quad \vee \quad \vee \quad \vee \quad \vee \quad \vee \quad \vee$

...

$\quad b_1 \quad b_2 \quad b_3 \quad b_4 \quad \cdots \quad b_{n-2} \quad b_{n-1}$

$b_{n-1} = a_n - a_{n-1}$

Adding up the corresponding sides of these equalities, we obtain

$$b_1 + b_2 + ... + b_{n-1} = a_n - a_1.$$

Therefore, for $n \geq 2$,

$$a_n = a_1 + (b_1 + b_2 + ... + b_{n-1})$$

$$= a_1 + \sum_{k=1}^{n-1} b_k.$$

Thus, for $n \geq 2$, a_n can be expressed in terms of the first term a_1 and the sum of the first $n-1$ terms of the progression of differences.

(**Demonstration 3**) Find the general term of the progression 1, 3, 7, 13, 21, 31, ...

[Solution] Take $\{a_n\}$ as the given progression and $\{b_n\}$ as the progression of differences.

$$2, 4, 6, 8, 10, ...$$

This is an arithmetic progression in which the first term is 2 and the common difference is 2. Therefore,

$$b_n = 2n.$$

Thus, for $n \geq 2$, a_n is

$$a_n = a_1 + \sum_{k=1}^{n-1} b_k = 1 + \sum_{k=1}^{n-1} 2k$$

$$= 1 + n(n-1) = n^2 - n + 1.$$

When we substitute $n = 1$ into the expression $n^2 - n + 1$, it becomes 1, which is the same as a_1. Therefore, the general term a_n we want to find is

$$a_n = n^2 - n + 1.$$

Problem 8 Find the general term of each of the following progressions. Then find the sum of the first n terms.

(1) 1, 2, 5, 10, 17, 26, ...

(2) 3, 4, 1, 10, -17, 64, ...

Demonstration 4 Find the following sum:

$$\frac{1}{1 \cdot 2} + \frac{1}{2 \cdot 3} + \frac{1}{3 \cdot 4} + \dots + \frac{1}{n(n + 1)} .$$

[**Solution**] In general, since $\frac{1}{k(k + 1)} = \frac{1}{k} - \frac{1}{k + 1}$,

$$\frac{1}{1 \cdot 2} + \frac{1}{2 \cdot 3} + \frac{1}{3 \cdot 4} + \dots + \frac{1}{n(n + 1)}$$

$$= (1 - \frac{1}{2}) + (\frac{1}{2} - \frac{1}{3}) + (\frac{1}{3} - \frac{1}{4}) + \dots + (\frac{1}{n} - \frac{1}{n + 1})$$

$$= 1 - \frac{1}{n + 1} = \frac{n}{n + 1} .$$

Problem 9 Find the following sums:

(1) $\displaystyle\sum_{k = 1}^{n} \frac{-1}{k(k + 2)}$

(2) $\displaystyle\sum_{k = 1}^{n} \frac{1}{(2k)^2 - 1}$

Problem 10 Use the identity

$$\frac{1}{k(k + 1)(k + 2)} = \frac{1}{2} \{ \frac{1}{k(k + 1)} - \frac{1}{(k + 1)(k + 2)} \}$$

to find $\displaystyle\sum_{k = 1}^{n} \frac{1}{k(k + 1)(k + 2)}$.

Demonstration 5   Find the general term of a progression in which the sum of the first n terms is $n^3 - n$.

[Solution]   Take a_n as the nth term of this progression and S_n as the sum of the first n terms.

Since for $n \geq 2$, $a_n = S_n - S_{n-1}$, we know that

$$a_n = (n^3 - n) - \{(n-1)^3 - (n-1)\}$$

$$= 3n(n-1). \tag{1}$$

For $n = 1$, since $a_1 = S_1$, we obtain

$$a_1 = 1^3 - 1 = 0. \tag{2}$$

From (1) and (2), for all natural numbers n

$$a_n = 3n(n-1).$$

Problem 11   Find the general term of a progression $\{a_n\}$ in which the sum S_n of the first n terms is $n^3 - n + 2$.

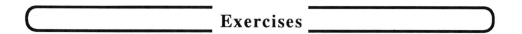

Exercises

1. Given a progression in which the first term is 5. For this progression the sum of the first 3 terms is equal to the sum of the first 5 terms. Find the common difference of the progression.

2. Given an arithmetic progression in which the 5th term is 108 and the 20th term is -237.

   (1) Find the first term and the common difference of this progression.

   (2) If the sum of the first n terms of this progression is a maximum, what is the value of n?

3. Given a geometric progression in which the third term is 12 and the 6th term is 96.

   (1) Find the sum of the squares of the first n terms.

   (2) Find the product of the first n terms.

4. Find the sum of the first n terms of the following progressions:

   (1) $1^2, 4^2, 7^2, 10^2, 13^2, ...$

   (2) $2 \cdot 3^2, 4 \cdot 4^2, 6 \cdot 5^2, 8 \cdot 6^2, 10 \cdot 7^2, ...$

5. Given a progression 2, 3, 9, 18, 28, 37, 43, 44, ... Find the general term of this progression by taking the progression of differences, and again taking the differences of the progression of differences.

6. Determine whether or not the progression in which the sum of the first n terms is expressed by a quadratic expression in n, $an^2 + bn + c$, is an arithmetic progression.

MATHEMATICAL INDUCTION
AND PROGRESSIONS

 Mathematical Induction

We can prove that for any natural number n,

$$1 + 3 + 5 + ... + (2n - 1) = n^2. \tag{1}$$

holds by the same method we used for finding the sum of an arithmetic progression. Here, let's consider another method of proving it.

Equality (1) is an expression for a natural number n; thus, we can express this number as $P(n)$ and prove the following two statements.

(I) $P(1)$ holds.

(II)   If we assume that $P(k)$ holds for any natural number k, then $P(k + 1)$ also holds.

[Proof]   (I) $P(1)$ means $1 = 1^2$, which clearly holds.

(II)   Assuming that $P(k)$ holds means assuming that the following equation holds:

$$1 + 3 + 5 + ... + (2k - 1) = k^2.$$

Adding $2k + 1$ to both sides of this equation, we obtain

$$1 + 3 + 5 + ... + (2k - 1) + (2k + 1) = k^2 + (2k +1).$$

Therefore,

$$1 + 3 + 5 + ... + (2k - 1) + (2k + 1) = (k + 1)^2.$$

Thus, $P(k + 1)$ holds.

We can take $k = 1, 2, 3, 4, \ldots$ in (II), and then we know that

if $P(1)$ holds, then $P(2)$ holds;

if $P(2)$ holds, then $P(3)$ holds;

if $P(3)$ holds, then $P(4)$ holds;

if $P(4)$ holds, then $P(5)$ holds;

$\ldots$.

If $P(1)$ holds, then

$$P(2), P(3), P(4), P(5), \ldots$$

all hold. We already know from (I) that $P(1)$ holds, and therefore $P(n)$ holds for all natural numbers n.

Thus, if we prove (I) and (II), then we can prove equation (1).

This method of proof is called **mathematical induction**.

Mathematical Induction

In order to prove that $P(n)$ holds for any natural number n, it is sufficient to prove (I) and (II):

(I) $P(n)$ holds for $n = 1$.

(II)   If we assume that $P(n)$ holds for $n = k$, then $P(n)$ also holds for $n = k + 1$.

Demonstration 1   Prove the following equality by mathematical induction, provided that n is a natural number.

$$1 \cdot 2 + 2 \cdot 3 + 3 \cdot 4 + \dots + n(n+1) = \frac{1}{3}n(n+1)(n+2) \qquad (1)$$

[Proof]

(I)   For $n = 1$,

left side $= 1 \cdot 2 = 2$, right side $= \frac{1}{3} \times 1 \times 2 \times 3 = 2$.

Therefore,   left side = right side.

(II)   Suppose that (1) holds for $n = k$. That is equivalent to supposing that

$$1 \cdot 2 + 2 \cdot 3 + \dots + k(k+1) = \frac{1}{3}k(k+1)(k+2).$$

Then, for $n = k + 1$,

left side

$$= 1 \cdot 2 + 2 \cdot 3 + \dots + k(k+1) + (k+1)(k+2)$$

$$= \frac{1}{3}k(k+1)(k+2) + (k+1)(k+2)$$

$$= \frac{1}{3}(k+1)(k+2)(k+3) = \text{right side.}$$

Therefore, (1) also holds for $n = k + 1$.

From (I) and (II), we know that equality (1) holds for all natural numbers.

Problem 1   Use mathematical induction to prove that the following equalities hold, provided that n is a natural number.

(1) $1^2 + 2^2 + \dots + n^2 = \dfrac{1}{6} n(n + 1)(2n + 1)$

(2) $\dfrac{1}{1 \cdot 2} + \dfrac{1}{2 \cdot 3} + \dots + \dfrac{1}{n(n + 1)} = \dfrac{n}{n + 1}$

(3) $1 \cdot 2 \cdot 3 + 2 \cdot 3 \cdot 4 + \dots + n(n + 1)(n + 2)$

$= \dfrac{1}{4} n(n + 1)(n + 2)(n + 3)$

Demonstration 2   For $h > 0$, use mathematical induction to prove that the following inequality holds for a natural number greater than or equal to 2.

$$(1 + h)^n > 1 + nh \qquad (1)$$

[Proof]

(I)   For $n = 2$, since $h^2 > 0$,

left side $= (1 + h)^2 = 1 + 2h + h^2 > 1 + 2h =$ right side.

(II)   For $k \geq 2$, suppose that (1) holds for $n = k$. That is equivalent to supposing that

$$(1 + h)^k > 1 + kh. \qquad (2)$$

Since $h > 0$, we know that $1 + h > 0$.

Multiplying both sides of (2) by $1 + h$, we obtain

$$(1 + h)^{k+1} > (1 + kh)(1 + h)$$

$$= 1 + (k + 1)h + kh^2 > 1 + (k + 1)h.$$

Therefore, (1) also holds for $n = k + 1$.

From (I) and (II), the given inequality (1) holds for all natural numbers greater than or equal to 2.

In the above proof, if we prove that the inequality holds for $n = 2$, as in (I), then we can use this result and (II) to prove that the inequality holds for all natural numbers greater than or equal to 2. This is also a proof by mathematical induction.

Problem 2   Given that $0 < a < b$, use mathematical induction to prove that $a^n < b^n$ holds for any natural number n.

Problem 3   Use mathematical induction to prove that the following inequality holds for all natural numbers greater than or equal to 2.

$$\frac{1}{1^2} + \frac{1}{2^2} + \frac{1}{3^2} + ... + \frac{1}{n^2} < 2 - \frac{1}{n}$$

Problem 4   Use mathematical induction to prove that the number of diagonals in a convex n-gon is $\dfrac{n(n-3)}{2}$.

Inductive Definition of a Progression

The following relation holds between the nth term a_n and the $(n+1)$th term a_{n+1} of an arithmetic progression with a common difference of d.

$$a_{n+1} = a_n + d \qquad (n = 1, 2, 3, ...)$$

Moreover, the following relation holds between the nth term a_n and the $(n+1)$th term a_{n+1} of a geometric progression with a common ratio of r.

$$a_{n+1} = a_n r \qquad (n = 1, 2, 3, ...)$$

A formula that expresses a_{n+1} in terms of a_n is called a **recurrence formula** for progression $\{a_n\}$.

For example, the recurrence formula of progression $\{a_n\}$ is given in the following way:

$$a_{n+1} = a_n + n \qquad (n = 1, 2, 3, ...).$$

The first term of this progression is 2. If we substitute 1, 2, 3, ... one by one for n in the above recurrence formula, we obtain:

$$a_2 = a_1 + 1 = 2 + 1 = 3;$$
$$a_3 = a_2 + 2 = 3 + 2 = 5;$$
$$a_4 = a_3 + 3 = 5 + 3 = 8,$$

... .

In this way, all the terms of the progression are decided one by one. So this progression $\{a_n\}$ can be defined as

$$\begin{cases} a_1 & = 2 \\ a_{n+1} = a_n + n \ (n = 1, 2, 3, \ldots) \end{cases}.$$

In general, if we have the first term a_1 and the recurrence formula expressing a_{n+1} in terms of a_n, we can find any term of a progression.

This method of defining a progression is generally called an **inductive definition**.

Problem 1 Progression $\{a_n\}$ is defined inductively by the following formulas. Find the 4th term.

(1) $\begin{cases} a_1 & = 3 \\ a_{n+1} = 2a_n - 4 \end{cases}$ 　　　　(2) $\begin{cases} a_1 & = 5 \\ a_{n+1} = 3a_n - 2n \end{cases}$

Problem 2 If progression $\{a_n\}$ is defined by the following formula, find a_7.

$$a_1 = 1, a_2 = 2$$

$$a_{n+2} = a_{n+1} + a_n \ (n = 1, 2, 3, \ldots)$$

Problem 3 Progression $\{a_n\}$ is defined by $a_1 = 1, a_{n+1} = pa_n + q$ $(n = 1, 2, 3, \ldots)$. Find the values of p and q, if the third term is 6 and the 5th term is 86.

Next, let's find the general term of a progression defined by a recurrence formula.

Demonstration 1　　　Find the general term of progression $\{a_n\}$, defined by $a_1 = 1, a_{n+1} = 1 + 4a_n$ $(n = 1, 2, 3, \ldots)$.

[Solution]　　Since $a_1 = 1$, we can find $a_2, a_3, \ldots$ one after another from the recurrence formula:

$$a_2 = 1 + 4;$$

$$a_3 = 1 + 4(1 + 4) = 1 + 4 + 4^2;$$

$$a_4 = 1 + 4(1 + 4 + 4^2) = 1 + 4 + 4^2 + 4^3;$$

$$\cdots.$$

Thus, we have

$$a_n = 1 + 4 + 4^2 + \ldots + 4^{n-1}.$$

Therefore,

$$a_n = 1 + 4 + 4^2 + \ldots + 4^{n-1} = \frac{4^n - 1}{4 - 1} = \frac{4^n - 1}{3}.$$

[Alternate Solution]   For $n \geq 2$,

$$a_{n+1} = 1 + 4a_n \, ;$$

$$a_n = 1 + 4a_{n-1}.$$

Subtracting both sides of the second equality from the corresponding sides of the first equality, we obtain

$$a_{n+1} - a_n = 4(a_n - a_{n-1}).$$

Then we set $b_n = a_{n+1} - a_n$ and rewrite the above expression as

$$b_n = 4b_{n-1}.$$

So progression $\{b_n\}$ is a geometric progression in which the common ratio is 4. And the first term of $\{b_n\}$ is

$$b_1 = a_2 - a_1 = 4;$$

$$b_n = 4 \times 4^{n-1} = 4^n.$$

Since progression $\{b_n\}$ is the progression of differences for progression $\{a_n\}$, we obtain the following formula for $n \geq 2$:

$$a_n = a_1 + \sum_{k=1}^{n-1} b_k = 1 + \sum_{k=1}^{n-1} 4^k$$

$$= 1 + 4 + 4^2 + \ldots + 4^{n-1}$$

$$= \frac{4^n - 1}{4 - 1} = \frac{4^n - 1}{3}.$$

This formula also holds for $n = 1$. Therefore, for all natural numbers n,

$$a_n = \frac{4^n - 1}{3}.$$

Problem 4 If we transform $a_{n+1} = 1 + 4a_n$, we get $a_{n+1} + \frac{1}{3} = 4\left(a_n + \frac{1}{3}\right)$.

Solve Demonstration 1 by setting $b_n = a_n + \frac{1}{3}$.

Problem 5 Find the general term of progression $\{a_n\}$ defined by the following formulas:

(1) $a_1 = 1$, $a_{n+1} = a_n + n$ $(n = 1, 2, 3, ...)$

(2) $a_1 = 1$, $3a_{n+1} = 2a_n + 3$ $(n = 1, 2, 3, ...)$

(3) $a_1 = 1$, $a_{n+1} = 4 + 3a_n$ $(n = 1, 2, 3, ...)$

Demonstration 2 There are n straight lines in a plane, no two of which are parallel, and no three of which intersect at a single point. Into how many parts do these lines divide the plane?

[Solution] Take a_n as the number of parts created by n lines. If we draw the $(n + 1)$th line, then this line intersects the original n lines at n points, and the number of parts created by the lines increases by $n + 1$. Therefore,

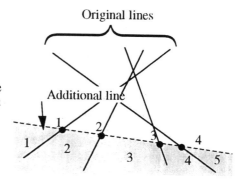

Original lines

Additional line

$$a_{n+1} = a_n + (n + 1). \tag{1}$$

Take $\{b_n\}$ as the progression of differences of progression $\{a_n\}$, and then from (1) we obtain

$$b_n = a_{n+1} - a_n = n + 1.$$

And since $a_1 = 2$, for $n \geq 2$,

$$a_n = a_1 + \sum_{k=1}^{n-1} b_k = 2 + \sum_{k=1}^{n-1} (k + 1).$$

Here,

$$\sum_{k=1}^{n-1} (k+1) = 2 + 3 + \dots + n = \frac{n(n+1)}{2} - 1.$$

Therefore,

$$a_n = 2 + \frac{n(n+1)}{2} - 1 = \frac{1}{2}(n^2 + n + 2).$$

If we set $n = 1$ in this formula, it becomes equal to 2 and coincides with a_1.

Thus, the plane is divided into $\frac{1}{2}(n^2 + n + 2)$ parts by n straight lines.

Problem 6   There are n circles in a plane, any two of which intersect at two points, and no three of which intersect at a single point. Into how many parts do these circles divide the plane?

———————— **Reference: The Binomial Theorem** ————————

The "binomial theorem" is originally studied as a part of "Statistics and Probability"; however, because of its great usefulness, let us give a simple explanation here.

You already know that the expansions of $(1 + x)^n$ for $n = 2$ and 3 are

$$(1 + x)^2 = 1 + 2x + x^2;$$

$$(1 + x)^3 = 1 + 3x + 3x^2 + x^3.$$

The expansion for $n = 4$ is

$$(1 + x)^4 = (1 + x)(1 + x)^3$$

$$= (1 + x)(1 + 3x + 3x^2 + x^3)$$

$$= 1 + 3x + 3x^2 + x^3 + x + 3x^2 + 3x^3 + x^4$$

$$= 1 + 4x + 6x^2 + 4x^3 + x^4.$$

As you can see from the above calculations, the sums of the coefficients of the adjacent terms in the expansion of $(1 + x)^3$ are the coefficients of x, x^2, and x^3 in the expansion of $(1 + x)^4$.

```
  1   3   3   1
   V   V   V
1   4   6   4   1
```

Problem 1 Expand $(1 + x)^5$ and check that the coefficients of $x, x^2, x^3,$ and x^4 represent the sums of the coefficients of the adjacent terms in the expansion of $(1 + x)^4$.

Including the special case of $n = 0$, if we arrange the coefficients in the expansion of $(1 + x)^n$ in the form of a triangle, the property we have just observed holds everywhere. This arrangement is called **Pascal's triangle**.

Problem 2 Use Pascal's triangle to expand $(1 + x)^6$.

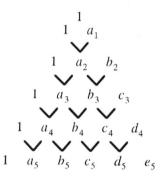

In general, if we write the coefficients of $(1 + x)^n$ as

$$(1 + x)^n = 1 + a_n x + b_n x^2 + c_n x^3 + \ldots + x^n, \tag{1}$$

then this property of Pascal's triangle tells us that:

$$a_{n+1} = a_n + 1;$$

$$b_{n+1} = b_n + a_n;$$

$$c_{n+1} = c_n + b_n.$$

Using these formulas, we can find formulas for the coefficients of expansion of each $(1 + x)^n$ successively.

Example 1 Since $a_1 = 1$ and $a_{n+1} = a_n + 1$, progression $\{a_n\}$ is an arithmetic progression in which the first term is 1 and the common difference is 1.

Therefore,

$$a_n = n.$$

Example 2   Since $a_n = n$, in general

$$b_{n+1} = b_n + n.$$

Moreover, since $b_2 = 1$,

$$b_2 = 1;$$

$$b_3 = b_2 + 2;$$

$$b_4 = b_3 + 3;$$

$$\dots$$

$$b_n = b_{n-1} + (n-1).$$

```
            1
          1   1
           \ /
          1  2  b_2
           \ / \ /
          1  3  b_3  c_3
           \ / \ / \ /
          1  4  b_4  c_4  d_4
           \ / \ / \ / \ /
          1  5  b_5  c_5  d_5  e_5
```

Thus,

$$b_n = 1 + 2 + 3 + \dots + (n-1) = \frac{n(n-1)}{2}.$$

Problem 3   Prove the following formula for coefficient c_n in (1):

$$c_n = \frac{n(n-1)(n-2)}{6}.$$

The constant term and the coefficients of x, x^2, ..., x^n in the expansion of $(1+x)^n$ can be expressed as follows, respectively:

$$_nC_0,\ _nC_1,\ _nC_2,\ \dots,\ _nC_n. \tag{2}$$

These expressions are called the **binomial coefficients**.

Since the constant term of the expansion of $(1+x)^n$ is 1,

$$_nC_0 = 1.$$

Moreover, from Examples 1, 2, and 3 we have

$$_nC_1 = n,\qquad _nC_2 = \frac{n(n-1)}{2},\qquad _nC_3 = \frac{n(n-1)(n-2)}{6}.$$

In general, the following formula holds for $r = 1, 2, ..., n$.

Binomial Coefficients

$$_nC_r = \frac{n(n-1)(n-2) \ldots (n-r+1)}{r!}$$

Here, $r!$ represents the product of the natural numbers from 1 through r $1 \cdot 2 \cdot 3 \cdot \ldots \cdot r$, and is called r **factorial**.

As you can see from Pascal's triangle, the numbers in the sequence of binomial coefficients are arranged symmetrically. Thus,

$$_nC_r = {}_nC_{n-r}.$$

Demonstration   Expand $(1 + x)^7$.

[Solution] $_7C_0 = 1,$ $_7C_1 = 7,$

$$_7C_2 = \frac{7 \cdot 6}{1 \cdot 2} = 21, \qquad _7C_3 = \frac{7 \cdot 6 \cdot 5}{1 \cdot 2 \cdot 3} = 35$$

Therefore,

$$(1 + x)^7 = 1 + 7x + 21x^2 + 35x^3 + 35x^4 + 21x^5 + 7x^6 + x^7.$$

Problem 4   Expand $(1 + x)^8$ by calculating the binomial coefficients.

Note: The primary meaning of the symbol $_nC_r$ is the total number of combinations created by drawing r objects from among n different objects.

Using the binomial coefficients, $(a + b)^n$ can be expanded in the following way:

$$(a + b)^n = a^n(1 + \frac{b}{a})^n$$

$$= a^n({}_nC_0 + {}_nC_1 \cdot \frac{b}{a} + {}_nC_2(\frac{b}{a})^2 + \dots + {}_nC_n(\frac{b}{a})^n)$$

$$= {}_nC_0a^n + {}_nC_1a^{n-1}b + {}_nC_2a^{n-2}b^2 + \dots + {}_nC_nb^n.$$

This is called the **binomial theorem**.

The Binomial Theorem

$$(a + b)^n = \sum_{r=0}^{n} {}_nC_r\, a^{n-r}b^r$$

$$= {}_nC_0a^n + {}_nC_1a^{n-1}b + {}_nC_2a^{n-2}b^2 + \dots + {}_nC_nb^n$$

Problem 5 Use the binomial theorem to expand the following expressions:

(1) $(a + b)^6$

(2) $(a - b)^6$

(3) $(2a + b)^5$

(4) $(2x - 3y)^4$

Exercises

1. Use mathematical induction to prove the following equalities, provided that n is a natural number:

    (1) $1^2 - 2^2 + 3^2 - 4^2 + ... - (2n)^2 = -n(2n + 1)$

    (2) $\dfrac{1}{2} \cdot \dfrac{3}{4} \cdot \dfrac{5}{6} \cdot ... \cdot \dfrac{2n - 1}{2n} < \sqrt{\dfrac{1}{2n + 1}}$

2. Use mathematical induction to prove that the following condition holds for any natural number n greater than or equal to 2:

    $"2^{3n} - 7n - 1$ is a multiple of 49."

3. Prove that the product of three consecutive natural numbers must be divisible by 6.

4. Progressions $\{a_n\}$ and $\{b_n\}$ are defined by $a_1 = 1$, $b_1 = 2$, $a_{n+1} = 2a_n + 3b_n$, and $b_{n+1} = 3a_n + 2b_n$. Find a_4 and a_5.

5. Find the general terms of the progressions defined by the following formulas:

    (1) $a_1 = 7$, $a_{n+1} - a_n = 4n$ $(n = 1, 2, 3, ...)$

    (2) $a_1 = -2$, $a_{n+1} - a_n = \dfrac{1}{1 + 2 + ... + n}$ $(n = 1, 2, 3, ...)$

    (3) $a_1 = 5$, $a_{n+1} = 2 + \dfrac{1}{3} a_n$ $(n = 1, 2, 3, ...)$

6. Progression $\{a_n\}$ is defined by $a_1 = 10$, and $2a_{n+1} = a_n + 6$ $(n = 1, 2, 3, ...)$. Find the sum S_n of the first n terms of this progression.

7. Find the general term of progression $\{a_n\}$ defined by $a_1 = 1$ and $a_{n+1} = 2^n a_n$ $(n = 1, 2, 3, ...)$.

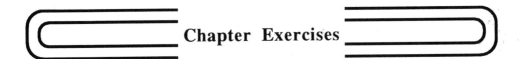

Chapter Exercises

A

1.　The sum of the first 3 terms is equal to the sum of the first 11 terms of an arithmetic progression in which the first term is 13. Find the common difference of this progression.

2.　$4, a, b$ and $b, c, 64$ are geometric progressions, and a, b, c is an arithmetic progression. Find the values of $a, b,$ and c.

3.　Find the sum of the first n terms of the following progressions:

　　(1)　$3 \cdot 1, \; 4 \cdot 4, \; 5 \cdot 7, \; 6 \cdot 10, \; ...$

　　(2)　$2, \; 22, \; 222, \; 2222, \; ...$

4.　Find the following sums:

　　(1)　$1 \cdot n + 2(n - 1) + 3(n - 2) + ... + (n - 1) \cdot 2 + n \cdot 1$

　　(2)　$1 + \dfrac{1}{1 + 2} + \dfrac{1}{1 + 2 + 3} + ... + \dfrac{1}{1 + 2 + 3 + ... + n}$

5.　Given the progression

　　$1, \; 1 + 2 + 1, \; 1 + 2 + 3 + 2 + 1, \; 1 + 2 + 3 + 4 + 3 + 2 + 1, \; ...$

　　(1)　Derive the general term by successively calculating each term of this progression.

　　(2)　Use mathematical induction to prove your derivation from (1).

6.　Progression $\{a_n\}$ is defined by $a_1 = 0, \; a_2 = 1, \; a_{n+2} = \dfrac{a_{n+1} + a_n}{2}$　$(n = 1, 2, 3, ...)$.

　　(1)　If we set $b_n = a_{n+1} - a_n$, what kind of progression is $\{b_n\}$?

　　(2)　Find the general term of progression $\{b_n\}$. Then find the general term of progression $\{a_n\}$.

B

1. Given the progression $2 \cdot 3,\ 4 \cdot 3^2,\ 6 \cdot 3^3,\ 8 \cdot 3^4,\ 10 \cdot 3^5,\ \dots.$

    (1) Find the general term.

    (2) Find the sum of the first n terms.

2. Progression $\{n\}$ is divided into groups by means of ( ) in the following way:

    $$(1),\ (2, 3),\ (4, 5, 6),\ (7, 8, 9, 10),\ \dots.$$

    The nth ( ) contains n numbers; express the first term inside those parentheses in terms of n. Then find the sum of the n terms inside the parentheses.

3. For progression $\{a_n\}$, the following equality holds for all natural numbers n :

    $$a_1 + 2a_2 + 3a_3 + \dots + na_n = n(n + 1)(n + 2).$$

    Find the general term of this progression.

4. Progression $\{a_n\}$ is defined by $a_1 = 1,\ a_{n+1} = \dfrac{a_n}{a_n + 2}$ $(n = 1, 2, 3, \dots).$

    (1) If we set $b_n = \dfrac{1}{a_n}$, find the relation between b_{n+1} and b_n.

    (2) Find the general term of progression $\{a_n\}$.

5. If n is 0 or a positive integer, take a_n as the number of paired integers (x, y) which satisfy

    $$y \le -\frac{x}{2} + n,\ x \ge 0,\ y \ge 0.$$

    (1) Find the relation between a_n and a_{n+1}.

    (2) Express a_n in terms of n.

6. Use mathematical induction to prove the following equalities, provided that n is a natural number.

    (1) $(n + 1)(n + 2) \dots (n + n) = 2^n \cdot 1 \cdot 3 \cdot 5 \cdot \dots \cdot (2n - 1)$

    (2) $1 - \dfrac{1}{2} + \dfrac{1}{3} - \dfrac{1}{4} + \dots - \dfrac{1}{2n} = \dfrac{1}{n + 1} + \dfrac{1}{n + 2} + \dots + \dfrac{1}{2n}$

CHAPTER 4

DIFFERENTIATION AND ITS APPLICATIONS

SECTION 1.   DIFFERENTIAL COEFFICIENTS AND DERIVATIVES
SECTION 2.   APPLICATIONS OF DIFFERENTIATION

Differentiation was developed in the seventeenth century in order to handle problems involving three different concepts: velocity, tangent lines, and maximum and minimum values. In its infancy, even the great scholars Galileo (1564-1642) and Descartes (1596-1650) experienced difficulty in performing calculations, made errors, and became confused when doing problems involving velocities. From the present day viewpoint, these were very simple exercises in differentiation and integration at the high school level. The same difficulties arose in dealing with tangent lines and maximum and minimum values.

It was only after the development of differentiation and integration in the seventeenth century that the concept of the function was introduced into mathematics.

The methods of performing differentiation and integration were perfected during the seventeenth century, although it took the whole century to work them out. The names of Newton (1642-1727) and Leibniz (1646-1716) should be recalled in this connection.

DIFFERENTIAL COEFFICIENTS AND DERIVATIVES

  ## Limits

In the quadratic function

$$y = x^2 + 2x - 3$$

as x comes closer and closer to 2 from either side in the graph at the right, y approaches infinitely close to 5.

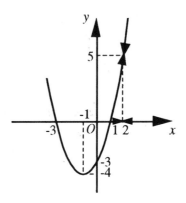

x	...	1.99	1.999	1.9999	...	2	...	2.0001	2.001	2.01	...
y	...	4.9401	4.9940...	4.9994...	...	5	...	5.0006...	5.0060...	5.0601	...

The function

$$y = f(x) = \frac{x^2 - 1}{x - 1}$$

is not defined at $x = 1$. However, at $x \neq 1$, it takes the form

$$f(x) = \frac{(x + 1)(x - 1)}{x - 1} = x + 1.$$

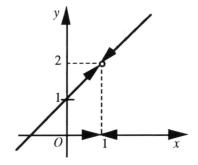

Therefore, as x approaches 1 without actually taking on the value of 1, $f(x)$ approaches infinitely close to 2.

In general, for a function $f(x)$, as x approaches closer and closer to a, while maintaining a value different from a, the value of $f(x)$ approaches a certain value b. We designate this as

$$f(x) \to b \text{ as } x \to a$$

or

$$\lim_{x \to a} f(x) = b.$$

b is called the **limit** of $f(x)$ as x approaches a.

Example
$$\lim_{x \to 2} (x^2 + 2x - 3) = 5, \qquad \lim_{x \to 1} \frac{x^2 - 1}{x - 1} = 2$$

Problem 1 Find the following limits:

(1) $\displaystyle \lim_{x \to 5} (4x - 7)$

(2) $\displaystyle \lim_{x \to -2} (2x^2 + 3x - 1)$

Demonstration 1 First find the value of $\displaystyle \lim_{x \to 3} (x^2 - 5)$ and $\displaystyle \lim_{x \to 3} (11 - x)$, and then find the following limits:

(1) $\displaystyle \lim_{x \to 3} \quad 4(x^2 - 5)$

(2) $\displaystyle \lim_{x \to 3} \quad \{(x^2 - 5) + (11 - x)\}$

(3) $\displaystyle \lim_{x \to 3} \quad (x^2 - 5)(11 - x)$

(4) $\displaystyle \lim_{x \to 3} \quad \frac{11 - x}{x^2 - 5}$

[Solution]

$$\lim_{x \to 3} (x^2 - 5) = 3^2 - 5 = 4$$

$$\lim_{x \to 3} (11 - x) = 11 - 3 = 8$$

Therefore,

(1) $\displaystyle \lim_{x \to 3} \quad 4(x^2 - 5) = 4 \times 4 = 16;$

(2) $\displaystyle \lim_{x \to 3} \quad \{(x^2 - 5) + (11 - x)\} = 4 + 8 = 12;$

(3) $\displaystyle \lim_{x \to 3} \quad (x^2 - 5)(11 - x) = 4 \times 8 = 32;$

(4) $\displaystyle \lim_{x \to 3} \quad \frac{11 - x}{x^2 - 5} = \frac{8}{4} = 2.$

The following formulas can be used to find the limits of functions.

Set $\lim_{x \to a} f(x) = \alpha$, $\lim_{x \to a} g(x) = \beta$ and let k be a constant.

(I) $\lim_{x \to a} kf(x) = k\alpha$

(II) $\lim_{x \to a} \{f(x) + g(x)\} = \alpha + \beta$

(III) $\lim_{x \to a} \{f(x) - g(x)\} = \alpha - \beta$

(IV) $\lim_{x \to a} f(x)g(x) = \alpha\beta$

(V) $\lim_{x \to a} \dfrac{f(x)}{g(x)} = \dfrac{\alpha}{\beta}$

(Provided that $\beta \neq 0$ in (V))

Problem 2   Find the following limits:

(1) $\lim_{x \to 1} \{(x^2 + x - 3) - (3x^2 - 1)\}$ (2) $\lim_{x \to -5} 3(x^2 - 2)$

(3) $\lim_{x \to 2} (x^3 + 1)(x - 3)^2$ (4) $\lim_{x \to -2} \dfrac{3x^2 - x - 4}{(2x + 7)^2}$

Demonstration 2   Find the following limits:

(1) $\lim_{x \to 1} \dfrac{2x^2 - 5x + 3}{x^2 + x - 2}$ (2) $\lim_{x \to 0} \dfrac{1}{x} \left(2 - \dfrac{4}{x + 2}\right)$

(3) $\lim_{x \to 1} \dfrac{\sqrt{x + 3} - 2}{x - 1}$

[Solution]   (1) $\lim_{x \to 1} \dfrac{2x^2 - 5x + 3}{x^2 + x - 2} = \lim_{x \to 1} \dfrac{(x - 1)(2x - 3)}{(x - 1)(x + 2)}$

$= \lim_{x \to 1} \dfrac{2x - 3}{x + 2} = -\dfrac{1}{3}$

(2) $\lim_{x \to 0} \dfrac{1}{x} \left(2 - \dfrac{4}{x + 2}\right) = \lim_{x \to 0} \dfrac{1}{x} \dfrac{2x + 4 - 4}{x + 2}$

$= \lim_{x \to 0} \dfrac{2}{x + 2} = \dfrac{2}{2} = 1$

(3) $\displaystyle\lim_{x \to 1} \frac{\sqrt{x + 3} - 2}{x - 1}$ $\displaystyle = \lim_{x \to 1} \frac{(x + 3) - 4}{(x - 1)(\sqrt{x + 3} + 2)}$

$\displaystyle = \lim_{x \to 1} \frac{1}{\sqrt{x + 3} + 2} = \frac{1}{\sqrt{4} + 2} = \frac{1}{4}$

Problem 3 Find the following limits:

(1) $\displaystyle\lim_{x \to -3} \frac{x^2 - 9}{x + 3}$

(2) $\displaystyle\lim_{x \to -2} \frac{x^2 + x - 2}{x^2 - 3x - 10}$

(3) $\displaystyle\lim_{x \to 0} \frac{1}{x} \left(\frac{1}{x + 1} - 1 \right)$

(4) $\displaystyle\lim_{x \to 2} \frac{x^2 - 4}{\sqrt{x + 2} - 2}$

Demonstration 3 Find the values of constants a and b such that the following equality holds:

$$\lim_{x \to 3} \frac{x^2 + ax + b}{x - 3} = 8.$$

[Solution] For $x \to 3$, the limit of the denominator is 0, and therefore, in order to satisfy the equality, we must have

$$x^2 + ax + b \to 0.$$

Thus,

$$\lim_{x \to 3} \frac{x^2 + ax + b}{x - 3} = 8.$$

Therefore,

$$\lim_{x \to 3} \frac{x^2 + ax + b}{x - 3} = \lim_{x \to 3} \frac{x^2 + ax - 3a - 9}{x - 3}$$

$$= \lim_{x \to 3} \frac{(x - 3)(x + a + 3)}{x - 3} = 6 + a$$

Since this limit is equal to 8,

$$6 + a = 8.$$

Therefore,

$$a = 2.$$

Thus,

$$b = -3 \times 2 - 9 = -15.$$

Answer: $a = 2$, $b = -15$

Problem 4   Find the values of constants a and b such that the following equalities hold:

(1) $\displaystyle \lim_{x \to 1} \frac{x^2 + ax + b}{x - 1} = 5$ (2) $\displaystyle \lim_{x \to -2} \frac{x^2 + ax - 6}{2x^2 + 3x - 2} = b$

  Differential Coefficients

Average Rate of Change

When an object falls freely, if we take s meters as the distance it has fallen after t seconds, then s can be expressed as a function of t:

$$s = 4.9t^2.$$

Using this formula, the average velocity at which it falls after between one and three seconds can be expressed as

$$\frac{4.9 \times 3^2 - 4.9 \times 1^2}{3 - 1} = 19.6 \ m/\text{sec}.$$

In general, if x varies from a to b in a function $y = f(x)$, the ratio of the change in the value of y, $f(b) - f(a)$, to the change in the value of x, $b - a$, is

$$\frac{f(b) - f(a)}{b - a}.$$

This ratio is called the **average rate of change** of the function $y = f(x)$ when x varies from a to b.

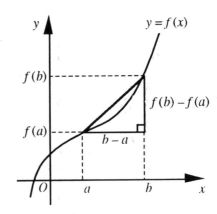

Demonstration 1   For the quadratic function $y = x^2$, find the average rate of change when x varies from a to b and when x varies from a to $a + h$.

[Solution]   The average rate of change from a to b is

$$\frac{b^2 - a^2}{b - a} = \frac{(b + a)(b - a)}{b - a} = b + a.$$

The average rate of change from a to $a + h$ is

$$\frac{(a + h)^2 - a^2}{(a + h) - a} = \frac{2ah + h^2}{h} = 2a + h.$$

Problem 1   For the linear function $y = px + q$, find the average rate of change from a to b.

Problem 2   (1)   For $f(x) = 2x^2 - 6x + 1$, find the average rate of change from -1 to 1.

(2)   For $f(x) = x^3 + 2$, find the average rate of change from a to $a + h$.

Differential Coefficients

Starting from the average rate of change of the quadratic function $y = x^2$ from a to b, if we take the limit as $b \to a$, then we obtain

$$\lim_{b \to a} \frac{f(b) - f(a)}{b - a} = \lim_{b \to a} \frac{b^2 - a^2}{b - a}$$

$$= \lim_{b \to a} (b + a) = 2a.$$

In general, given the average rate of change of a function $y = f(x)$ from a to x, as x approaches closer and closer to the fixed value a, if the limit

$$\lim_{x \to a} \frac{f(x) - f(a)}{x - a}$$

is defined, then we call this value the **differential coefficient** or **rate of change** of the function $y = f(x)$ at $x = a$, and it is designated by symbol $f'(a)$.

If we take $x - a = h$, then this expression takes the following form:

Differential Coefficients

The differential coefficient $f'(a)$ of a function $y = f(x)$ at $x = a$ is

$$f'(a) = \lim_{h \to 0} \frac{f(a + h) - f(a)}{h}.$$

Demonstration 2   Given the function $f(x) = x^2 + 2x - 4$, find $f'(3)$.

[Solution] $f'(3) = \lim_{h \to 0} \dfrac{f(3 + h) - f(3)}{h}$

$= \lim_{h \to 0} \dfrac{\{(3 + h)^2 + 2(3 + h) - 4\} - (3^2 + 2 \times 3 - 4)}{h}$

$= \lim_{h \to 0} \dfrac{6h + h^2 + 2h}{h}$

$= \lim_{h \to 0} (8 + h) = 8$

Problem 3   Given the function $f(x)$ from Demonstration 2, find $f'(0)$ and $f'(-2)$.

Problem 4   Find the differential coefficient of the function for the indicated value of x:

(1) $f(x) = 5 - 3x^2$ (-3)

(2) $f(x) = x^3 - 2x$ (1)

Problem 5   The average rate of change of the quadratic function $f(x) = px^2 + qx + r$ from $x = a$ to $x = b$ is equal to the differential coefficient at $x = c$. Find the relation among a, b, and c.

Derivatives

Given a function $y = f(x)$, if we relate each value a of x to the differential coefficient $f'(a)$, then we obtain a new function. This new function is designated by $f'(x)$ and is called the **derivative** of the function $f(x)$.

$$f'(x) = \lim_{h \to 0} \frac{f(x + h) - f(x)}{h}$$

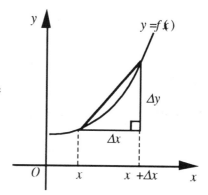

The expression $f(x + h) - f(x)$ represents the change in the value of y according to h, the change in the value of x. The changes in the value of x and y are called the **increment** of x and the **increment** of y, and are designated by Δx and Δy.*

Thus,

$\Delta x = h$,

$\Delta y = f(x + h) - f(x) = f(x + \Delta x) - f(x)$.

Using this symbol, the derivative can be expressed in the following form:

Derivatives

$$f'(x) = \lim_{\Delta x \to 0} \frac{\Delta y}{\Delta x} = \lim_{\Delta x \to 0} \frac{f(x + \Delta x) - f(x)}{\Delta x}$$

Finding the derivative $f'(x)$ of a function $f(x)$ is referred to as **differentiating** $f(x)$ with respect to x.

The following symbols are also used to express derivatives:

$$y', \qquad \frac{dy}{dx}, \qquad \frac{d}{dx} f(x).$$

The value of a derivative $f'(x)$ at $x = a$ is clearly the same as $f'(a)$, the differential coefficient of $f(x)$ at $x = a$.

---

* Δ is the Greek letter delta corresponding to D, the first letter of *Difference*.

(Demonstration 1)   Find the derivative of the function $y = x^2 + 5x - 2$.

[Solution] $\Delta y = \{(x + \Delta x)^2 + 5(x + \Delta x) - 2\} - (x^2 + 5x - 2)$

$$= 2x\Delta x + 5\Delta x + (\Delta x)^2$$

Therefore,

$$\frac{dy}{dx} = \lim_{\Delta x \to 0} \frac{\Delta y}{\Delta x} = \lim_{\Delta x \to 0} (2x + 5 + \Delta x)$$

$$= 2x + 5.$$

(Problem 1)   Find the derivatives of the following functions:

(1) $y = 3x^2 - 7$ (2) $y = x^3 + 4x - 3$

Calculating Derivatives

If we use the definition of a derivative to find the derivatives of the functions

$$f(x) = x^2, \ f(x) = x^3, \ f(x) = x^4,$$

we obtain:

$$f'(x) = 2x, \ f'(x) = 3x^2, \ f'(x) = 4x^3.$$

(Problem 2)   Check that the above statement is a fact.

We can formulate the following generalization:

The Derivative of x^n

For a positive integer n,

if $f(x) = x^n$, then $f'(x) = nx^{n-1}$.

[Proof]   Find the limit of

$$\frac{f(x + \Delta x) - f(x)}{\Delta x} = \frac{(x + \Delta x)^n - x^n}{\Delta x} \tag{1}$$

as Δx approaches 0.

If we set $u = x + \Delta x$, then the numerator of the right side can be transformed into

$$u^n - x^n$$

$$= (u^n - u^{n-1}x) + (u^{n-1}x - u^{n-2}x^2) + \dots + (ux^{n-1} - x^n)$$

$$= u^{n-1}(u - x) + u^{n-2}x(u - x) + \dots + x^{n-1}(u - x)$$

$$= (u - x)(u^{n-1} + u^{n-2}x + \dots + x^{n-1}).$$

Therefore, from $u - x = \Delta x$,

$$\frac{f(x + \Delta x) - f(x)}{\Delta x}.$$

$$= u^{n-1} + u^{n-2}x + \dots + ux^{n-2} + x^{n-1}. \tag{2}$$

Since $u \to x$, as $\Delta x \to 0$, all n terms of the right side approach x^{n-1}.

Thus,

$$f'(x) = \lim_{\Delta x \to 0} \frac{f(x + \Delta x) - f(x)}{\Delta x} = nx^{n-1}.$$

Problem 3   Find the derivatives of the following functions:

(1) $y = x^6$

(2) $y = x^7$

The following formulas can be used for finding the derivatives of the sum and difference of two functions.

> Here, $f(x)$ and $g(x)$ are given functions, and c and k are constants.
>
> [I]   If $y = c$, then $y' = 0$.
>
> [II]   If $y = kf(x)$, then $y' = kf'(x)$.
>
> [III]   If $y = f(x) + g(x)$, then $y' = f'(x) + g'(x)$.
>
> [IV]   If $y = f(x) - g(x)$, then $y' = f'(x) - g'(x)$.

[Proof]

[I] $\Delta y = f(x + \Delta x) - f(x) = c - c = 0$

Therefore,

$$y' = \lim_{\Delta x \to 0} \frac{\Delta y}{\Delta x} = \lim_{\Delta x \to 0} \frac{0}{\Delta x} = 0.$$

[II] $\Delta y = kf(x + \Delta x) - kf(x) = k\{f(x + \Delta x) - f(x)\}$

Therefore,

$$y' = \lim_{\Delta x \to 0} \frac{\Delta y}{\Delta x} = \lim_{\Delta x \to 0} \frac{k\{f(x + \Delta x) - f(x)\}}{\Delta x}.$$

$$= k \lim_{\Delta x \to 0} \frac{f(x + \Delta x) - f(x)}{\Delta x} = kf'(x).$$

[III] $\Delta y = \{f(x + \Delta x) + g(x + \Delta x)\} - \{f(x) + g(x)\}$

$$= \{f(x + \Delta x) - f(x)\} + \{g(x + \Delta x) - g(x)\}$$

Therefore,

$$y' = \lim_{\Delta x \to 0} \frac{\Delta y}{\Delta x}$$

$$= \lim_{\Delta x \to 0} \{\frac{f(x + \Delta x) - f(x)}{\Delta x} + \frac{g(x + \Delta x) - g(x)}{\Delta x}\}$$

$$= \lim_{\Delta x \to 0} \frac{f(x + \Delta x) - f(x)}{\Delta x} + \lim_{\Delta x \to 0} \frac{g(x + \Delta x) - g(x)}{\Delta x}$$

$$= f'(x) + g'(x).$$

[IV]   can be proved analogously to [III].

Problem 4   Prove that if $y = f(x) + g(x) + h(x)$, then the derivative $y' = f'(x) + g'(x) + h'(x)$.

Demonstration 2   Differentiate the following functions:

(1) $y = 7x^8$

(2) $y = 3x^2 - 2x + 4$

(3) $y = (x + 1)(x - 1)$

[Solution]   (1) $y' = (7x^8)' = 7(x^8)' = 7 \times 8x^7 = 56x^7$

(2) $y' = (3x^2 - 2x + 4)' = (3x^2)' - (2x)' + (4)'$

$= 6x - 2$

(3)   Since $y = x^2 - 1$,

$y' = (x^2 - 1)' = (x^2)' - (1)' = 2x.$

Problem 5   Differentiate the following functions:

(1) $y = 4x^2 + 7x - 1$ (2) $y = 3 + 6x^2 - 7x^4$

(3) $y = (x + 1)(3x - 2)$ (4) $y = x(x - 1)(x - 3)$

Problem 6   (1)   Check that if $y = (ax + b)^2$, then $y' = 2a(ax + b)$.

(2)   Check that if $y = (ax + b)^3$, then $y' = 3a(ax + b)^2$.

Problem 7   Given the cubic function $f(x) = ax^3 + bx^2 + cx - 4$. Find values for the constants a, b, and c such that $f(2) = 0$, $f'(1) = 0$, and $f'(-2) = 0$.

Exercises

1. Find the following limits:

(1) $\lim\limits_{x \to 3} (x^2 - 3x + 4)$

(2) $\lim\limits_{x \to -2} (x^3 - 2x + 8)$

(3) $\lim\limits_{x \to 4} \dfrac{x + 2}{x^2 + x - 18}$

(4) $\lim\limits_{x \to -1} \dfrac{x^2 + 3x + 2}{x^2 - 2x - 3}$

(5) $\lim\limits_{x \to 0} \dfrac{5x + x^3}{4x - x^2}$

(6) $\lim\limits_{x \to 1/2} \dfrac{2x^2 - x}{2x^2 - 3x + 1}$

2. Find the values of the constants a and b such that the following equality holds:

$$\lim_{x \to 2} \frac{x^2 - ax + 8}{x^2 - (b + 2)x + 2b} = \frac{1}{5}.$$

3. Find the cubic function $f(x)$ which satisfies $\lim\limits_{x \to 0} \dfrac{f(x)}{x} = 2$ and $\lim\limits_{x \to 1} \dfrac{f(x)}{x - 1} = 1$.

4. The slope of the straight line connecting two points $(1, f(1))$ and $(a, f(a))$ on the graph of $f(x) = x^3$ coincides with the rate of change at $x = p$. Express p in terms of a, provided that $1 < p < a$.

5. Differentiate the following functions:

(1) $y = x^5 - 6x^3 + 3x$

(2) $y = x^2(7 - 3x^2)$

(3) $y = (4x - 1)(3x + 2)$

(4) $y = (5x - 1)^2$

6. Solve the following problems, assuming that a is a constant.

(1) When the radius r of a circle varies, find the differential coefficient of the area S of this circle for $r = a$.

(2) When the radius r of a ball varies, find the differential coefficient of the volume V of this ball for $r = a$.

APPLICATIONS OF DIFFERENTIATION

 Tangent Lines

Let's consider the meaning of the differential coefficient of a function $y = f(x)$ at $x = x_1$ for the graph of the function.

The definition of the differential coefficient is

$$f'(x_1) = \lim_{x \to x_1} \frac{f(x) - f(x_1)}{x - x_1}.$$

Here,

$$\frac{f(x) - f(x_1)}{x - x_1}$$

is the average rate of change, and it represents the slope of the straight line connecting the two points $P(x_1, f(x_1))$ and $Q(x, f(x))$ on the graph.

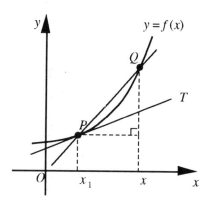

Next, if $x \to x_1$, then point Q moves along the graph and approaches point P, and line PQ approaches the fixed line PT through point P with a slope of $f'(x_1)$. This fixed line PT is called the **tangent line** of the function $y = f(x)$ at point P on the graph, and point P is called the **tangent point**.

The differential coefficient $f'(x_1)$ represents the slope of tangent line PT. Therefore, the equation of the tangent line has the following form:

Equation of a Tangent Line

The equation of the tangent line to a function $y = f(x)$ at the point $(x_1, f(x_1))$ is

$$y - f(x_1) = f'(x_1)(x - x_1).$$

Demonstration 1   Given the graph of function $y = x^3 - 7x + 5$.

(1) Find the equation of the tangent line at the point (2, -1) on the graph.

(2) Find the equation of the tangent line with a slope of - 4.

[Solution]   (1)   Take $f(x) = x^3 - 7x + 5$. Then

$$f'(x) = 3x^2 - 7.$$

Therefore,
$$f'(2) = 12 - 7 = 5.$$

Thus, the tangent line we want to find is a line through the point (2, -1) with a slope of 5. Therefore, the equation is

$$y + 1 = 5(x - 2) \quad \text{or} \quad y = 5x - 11.$$

(2) The slope of the tangent line at the point (x_1, y_1) on the graph is $f'(x_1)$, and the value is now - 4. Therefore,

$$f'(x_1) = 3x_1^2 - 7 = -4.$$

Thus, $x_1 = \pm 1.$

$$\text{For } x_1 = 1, \quad y_1 = -1.$$

$$\text{For } x_1 = -1, \quad y_1 = 11.$$

Therefore, there are two tangent points of tangent lines with a slope of - 4: (1, -1) and (-1, 11).

Thus, the equations of the tangent lines are

$$y + 1 = -4(x - 1) \quad \text{or} \quad y = -4x + 3;$$

$$y - 11 = -4(x + 1) \quad \text{or} \quad y = -4x + 7.$$

Problem 1   Find the equations of tangent lines at a point with an x -coordinate of 2 on the graphs of the following functions:

(1) $y = -2x^2 + 10$ (2) $y = 5 + x - x^3$

Problem 2 Find the equation of a straight line parallel to the line $y = 3x + 1$ and tangent to the graph of $y = -x^3 + 3x + 4$.

Problem 3 Find the coordinates of the point at which the tangent line of the curve $y = x^3 - 4x$ at the point $(1, -3)$ on this curve intersects the curve a second time.

Demonstration 2 Find the equation of the tangent line of the parabola $y = x^2 - 3x$ through point $A(3, -4)$.

[Solution] Take the tangent point as point $P(\alpha, \alpha^2 - 3\alpha)$. Since $y' = 2x - 3$, the equation of the tangent line AP is

$$y - (\alpha^2 - 3\alpha) = (2\alpha - 3)(x - \alpha).$$

Thus,

$$y = (2\alpha - 3)x - \alpha^2. \qquad (1)$$

Since this tangent line passes through the point $A(3, -4)$,

$$-4 = 3(2\alpha - 3) - \alpha^2.$$

$$\alpha^2 - 6\alpha + 5 = 0, \quad (\alpha - 1)(\alpha - 5) = 0$$

Therefore,

$$\alpha = 1, 5.$$

Substituting these values into (1), we obtain the following equations for the tangent lines:

$$y = -x - 1, \quad y = 7x - 25.$$

Problem 4 Find the equation of the tangent line from point $(1, 5)$ to the curve $y = x^3$.

Problem 5 Draw two lines tangent to the parabola $y = x^2 + 1$ from the point (-1, -7). Find the coordinates of the tangent points.

Increasing and Decreasing Functions

Increasing and Decreasing Functions

Given two real numbers a and b, sets of real numbers x satisfying inequalities such as

$$a < x < b, \qquad a \le x \le b,$$

$$a < x \le b, \qquad a \le x < b,$$

are called **intervals**. Sets of real numbers satisfying inequalities such as

$$x < a, \qquad b \le x,$$

are also called intervals.

Over a particular interval, if the value of a function $f(x)$ increases as the value of the variable x increases, $f(x)$ is said to be **increasing** on that interval, and is called an **increasing function**.

If the value of $f(x)$ decreases as the value of the variable x increases, $f(x)$ is said to be **decreasing** on that interval, and is called a **decreasing function**.

In other words, for any two numbers x_1 and x_2 in an interval, if the implication

$$x_1 < x_2 \Rightarrow f(x_1) < f(x_2)$$

holds, $f(x)$ is an increasing function. And if the implication

$$x_1 < x_1 \Rightarrow f(x_1) > f(x_2)$$

holds, $f(x)$ is a decreasing function.

Example 1 (1) Let $y = f(x) = x^2$.

For any two numbers x_1 and x_2 such that $0 < x_1 < x_2$ we have

$$f(x_1) - f(x_2)$$

$$= x_1^2 - x_2^2$$

$$= (x_1 - x_2)(x_1 + x_2)$$

$$< 0.$$

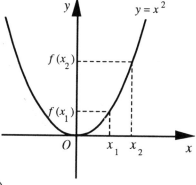

Therefore, $f(x_1) < f(x_2)$.

In the same way, for any two numbers x_1 and x_2 such that $x_1 < x_2 < 0$ we have

$$f(x_1) > f(x_2).$$

Therefore, the function $y = x^2$ increases on the interval $0 < x$, and decreases on the interval $x < 0$.

(2) Let $y = f(x) = x^3$.

For any two numbers x_1 and x_2 such that $x_1 < x_2$ we have

$$f(x_1) - f(x_2)$$

$$= x_1^3 - x_2^3$$

$$= (x_1 - x_2)(x_1^2 + x_1 x_2 + x_2^2)$$

$$< 0.$$

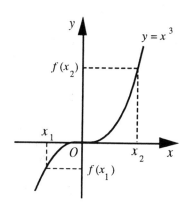

Therefore, $f(x_1) < f(x_2)$.

Thus, the function $y = x^3$ always increases.

Problem 1   Determine whether the following functions are increasing or decreasing on the indicated interval:

(1) $y = 4 - x^2$ $(x < 0)$

(2) $y = x^4$ $(x > 0)$

Derivatives and Increasing and Decreasing Functions

The differential coefficient $f'(x_1)$ of the function

$$y = f(x)$$

at $x = x_1$ is the limit of

$$\frac{f(x) - f(x_1)}{x - x_1}$$

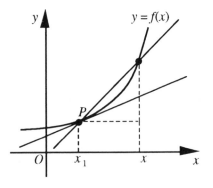

as $x \to x_1$. Therefore, if

$$f'(x_1) > 0,$$

then when x is very close to x_1,

$$\frac{f(x) - f(x_1)}{x - x_1} > 0.$$

Thus,

for $x > x_1$: $f(x) > f(x_1)$;

for $x < x_1$: $f(x) < f(x_1)$.

Therefore, $f(x)$ is increasing near $x = x_1$.

Analogously, if

$$f'(x_1) < 0,$$

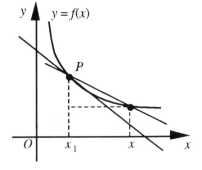

then $f(x)$ is decreasing near $x = x_1$.

On a particular interval:

if $f'(x) > 0$ always holds, then $f(x)$ is increasing on that interval;

if $f'(x) < 0$ always holds, then $f(x)$ is decreasing on that interval.

If $f'(x) = 0$ always holds on an interval, $f(x)$ is a constant on that interval.

Example 2 (1) Since the derivative of the function

$$f(x) = x^2 - 6x + 1$$

is

$$f'(x) = 2(x - 3),$$

we see that

for $x < 3$, $f'(x) < 0$;

for $x > 3$, $f'(x) > 0$.

x	$x < 3$	3	$3 < x$
$f'(x)$	–	0	+
$f(x)$	decr.	-8	incr.

Thus, $f(x)$ is decreasing for $x < 3$ and increasing for $3 < x$.

(2) Since the derivative of the function

$$f(x) = x^3 + 4x - 7$$

is

$$f'(x) = 3x^2 + 4$$

and $f'(x) > 0$ always holds, $f(x)$ is always increasing.

（Demonstration 1）   Determine when the function $y = x^3 + 3x^2 - 9x - 10$ is increasing and decreasing.

[Solution] $y' = 3x^2 + 6x - 9 = 3(x + 3)(x - 1)$

Therefore,

for $x < -3, \ y' > 0$;

for $-3 < x < 1, \ y' < 0$;

for $1 < x, \ y' > 0$.

Thus, y is increasing for $x < -3$, decreasing for $-3 < x < 1$, and increasing for $1 < x$.

（Problem 2）   Determine when the following functions are increasing and decreasing:

(1) $y = 12x - x^3$ (2) $y = x^4 - 6x^2 + 1$

Local Maximum and Local Minimum of a Function

We consider when the function

$$f(x) = x^3 + 3x^2 - 9x - 10$$

is increasing and decreasing in Demonstration 1. As x increases and passes -3, $f(x)$ changes from increasing to decreasing.

x	$x < -3$	-3	$-3 < x < 1$	1	$1 < x$
$f'(x)$	+	0	−	0	+
$f(x)$	incr.	17	decr.	-15	incr.

Therefore, $f(-3)$ is the maximum value of $f(x)$ near $x = -3$. In this case, $f(x)$ is said to have a **local maximum** at $x = -3$, and $f(-3)$, the value of $f(x)$ at $x = -3$, is the **local maximum value.**

When x passes $x = 1$, $f(x)$ changes from decreasing to increasing.

Thus, $f(1)$ is the minimum value of $f(x)$ near $x = 1$. In this case, the function $f(x)$ is said to have a **local minimum** at $x = 1$, and $f(1)$, the value of $f(x)$ at $x = 1$, is the **local minimum value**.

The local maximum value and the local minimum value, taken together, are called **extreme values**.

We can formulate the following generalization.

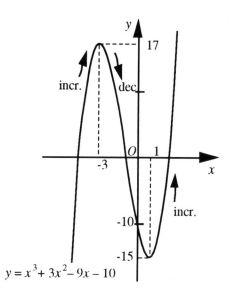

$$y = x^3 + 3x^2 - 9x - 10$$

The Local Maximum and Local Minimum of a Function

As x increases and passes x_1:

(1) if the sign of $f'(x)$ changes from plus to minus, $f(x)$ has a local maximum at x_1 and $f'(x_1)$ is the local maximum value;

(2) if the sign of $f'(x)$ changes from minus to plus, $f(x)$ has a local minimum at x_1 and $f(x_1)$ is the local minimum value.

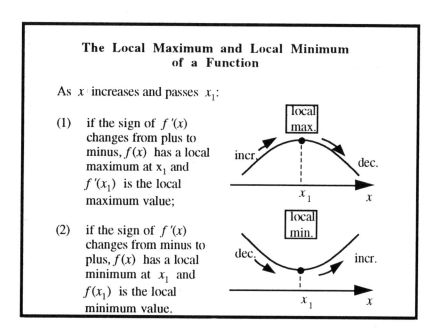

(Demonstration 2)   Determine when the function $f(x) = x^3 + 3x^2 - 2$ is increasing and decreasing, and find the extreme values.

[Solution] $f'(x) = 3x^2 + 6x = 3x(x + 2)$

Let's consider the sign of $f'(x)$ and whether $f(x)$ is increasing or decreasing, and then compile a table to show when it is increasing and decreasing.*

x	$x < -2$	-2	$-2 < x < 0$	0	$0 < x$
$f'(x)$	+	0	–	0	+
$f(x)$	↗	local max 2	↘	local min -2	↗

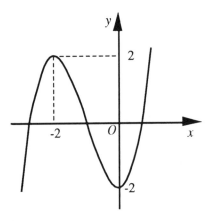

From this table, we can see that the function is:

increasing for $x < -2$,

decreasing for $-2 < x < 0$,

increasing for $0 < x$,

it has a local maximum at $x = -2$, and the local maximum value is 2;

it has a local minimum at $x = 0$, and the local minimum value is -2.

---

* In the table, ↗ means increasing and ↘ means decreasing.

Problem 3   Determine when the following functions are increasing and decreasing, and find their extremes:

(1) $y = 3x - x^3$ (2) $y = x^3 + 6x^2 - 15x + 4$

(3) $y = \dfrac{1}{4}x^4 - \dfrac{1}{2}x^3 - x^2$

Finding the Extremes

In order to find the extremes of a function $f(x)$, it is sufficient to find the values of x such that

$$f'(x) = 0$$

and examine the sign of $f'(x)$ near this value of x.

Demonstration 3   Determine when the function $f(x) = x^3(x + 2)$ is increasing and decreasing, and find its extremes. Then draw the graph.

[Solution]   Differentiating $f(x) = x^4 + 2x^3$, we obtain

$$f'(x) = 4x^3 + 6x^2 = 2x^2(2x + 3).$$

The solutions for $f'(x) = 0$ are

$$x = 0, \ -\frac{3}{2} \ .$$

When we examine the sign of $f'(x)$ near these values and determine whether $f(x)$ is increasing or decreasing, we obtain the following table:

x	$x < -\dfrac{3}{2}$	$-\dfrac{3}{2}$	$-\dfrac{3}{2} < x < 0$	0	$0 < x$
$f'(x)$	$-$	0	$+$	0	$+$
$f(x)$	↘	local min	↗	0	↗

Therefore, this function has a local minimum at $x = -\dfrac{3}{2}$, and the local minimum value is

$$f(-\tfrac{3}{2}) = -\frac{27}{16}.$$

If we graph this function using the additional facts that the graph intersects the x-axis at $(0, 0)$ and $(-2, 0)$, and that the graph is tangent to the x-axis at $x = 0$, then we obtain the graph in the figure to the right.

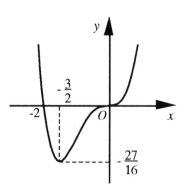

As you know from Demonstration 3, even if $f'(x_1) = 0$, if the sign of $f'(x_1)$ does not change before and after x_1, $f(x)$ does not have an extreme value at x_1.

Problem 4   Determine when the following functions are increasing and decreasing, and find their extremes. Then draw the graphs.

(1) $y = x^4 - 4x^3 + 1$ (2) $y = x^2(x - 1)(x + 1)$

(3) $y = x^4 - 6x^2 + 8x + 9$ (4) $y = x^6 - 3x^2$

Problem 5   Determine when the following functions are increasing and decreasing, and demonstrate that they have no extremes.

(1) $y = x^3 - 3x^2 + 3x + 3$ (2) $y = -x^5 - 5x^3 - 10x$

Problem 6   For what range of k will the function $y = x^3 + 3kx^2 - kx - 1$ have neither a local maximum value nor a local minimum value?

Maximum and Minimum Values

The method of using the sign of the derivative to determine when functions are increasing and decreasing and to find their extremes can be applied to the problem of finding the maximum and minimum values of functions.

Demonstration 4

We have a square piece of construction paper with sides of 10 cm.

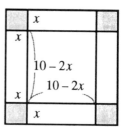

We make a box without a lid by cutting identical squares of side x cm from the four corners and folding the remaining paper. We want to make the volume of this box as great as possible. How long should we make the sides of the squares cut from the four corners?

[Solution]

If we take $f(x)$ cm^3 as the volume of the box, we have

$$f(x) = x(10 - 2x)^2 = 4x^3 - 40x^2 + 100x. \qquad (1)$$

Since $x > 0$ and $10 - 2x > 0$, the range of x is $0 < x < 5$.

Differentiating (1), we obtain

$$f'(x) = 12x^2 - 80x + 100 = 4(3x^2 - 20x + 25)$$

$$= 4(3x - 5)(x - 5).$$

Since $0 < x < 5$, we can determine when $f(x)$ is increasing and decreasing on the interval $0 < x < 5$ and compile the table below.

x	0	$0 < x < \dfrac{5}{3}$	$\dfrac{5}{3}$	$\dfrac{5}{3} < x < 5$	5
$f'(x)$		+	0	−	
$f(x)$		↗	local max.	↘	

If we express this information in a graph, we obtain the figure to the right.

Therefore, the function $f(x)$ has a local maximum at $x = \dfrac{5}{3}$. That means that we should make the sides of the square we cut out $\dfrac{5}{3}$ cm long.

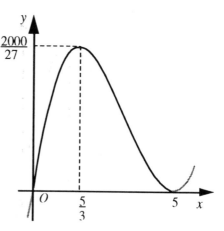

Problem 7   Find the maximum and minimum values of the function

$f(x) = x^3 - 3x^2 + 2$ on the following intervals.

(1) $1 \leq x \leq 3$ (2) $-2 \leq x \leq 4$

Problem 8   We have a right cylinder in which the sum of the height and the radius of the base is 30 cm. Find the height and the radius of the base when the volume is a maximum.

   Application of Differentiation to Equations and Inequalities

By considering the graph of the function $f(x)$, we can determine the number of solutions to the equation $f(x) = 0$, and we can prove that an inequality such as $f(x) > 0$ holds on a certain interval.

Demonstration 1   Determine the number of different real solutions to the cubic equation $x^3 - 3x^2 - a = 0$.

[Solution]   Transforming

$x^3 - 3x^2 - a = 0$,

we obtain

x	$x < 0$	0	$0 < x < 2$	2	$2 < x$
$f'(x)$	$+$	0	$-$	0	$+$
$f(x)$	↗	local max.	↘	local min.	↗

$x^3 - 3x^2 = a$.   (1)

Set

$f(x) = x^3 - 3x^2$.

Then,

$f'(x) = 3x^2 - 6x$

$= 3x(x - 2)$.

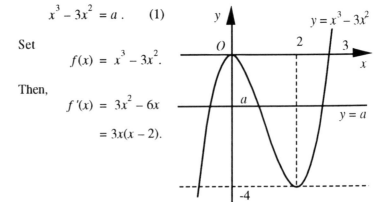

Therefore, we obtain the table above showing when the function is increasing and decreasing, and the local maximum value is $f(0) = 0$;

the local minimum value is $f(2) = -4$.

Thus, the graph takes the form shown above.

The number of real solutions is the number of points at which the graph of $y = f(x)$ intersects the straight line $y = a$. Therefore, the number of different real solutions is

for $a < -4$ or $a > 0$:   1;

for $a = -4$ or $a = 0$:   2;

for $-4 < a < 0$:   3.

Problem 1   Determine the range of the real number p such that the cubic equation $2x^3 - 3x^2 - 12x + p = 0$ has the following solutions:

(1)   three different real solutions;

(2)   two different positive solutions and one negative solution.

Problem 2   The cubic equation $x^3 - 3x + a = 0$ has only one real solution, and that solution is positive. Find the range of the real number a.

Demonstration 2   Prove that the inequality $3x^4 + 1 > 8x^3 - 18x^2$ holds for $x \geq 0$.

[Proof]   Set $f(x) = 3x^4 - 8x^3 + 18x^2 + 1$. We have

$$f'(x) = 12x^3 - 24x^2 + 36x$$

$$= 12x(x^2 - 2x + 3).$$

Since $x^2 - 2x + 3 = (x - 1)^2 + 2 > 0$,

for $x > 0$, $f'(x) > 0$.

Therefore, $f(x)$ is increasing for $x > 0$.

Moreover,

$$f(0) = 1 > 0.$$

Thus,

$$\text{for } x \geq 0, \ f(x) = 3x^4 - 8x^3 + 18x^2 + 1 > 0.$$

Accordingly,

$$3x^4 + 1 > 8x^3 - 18x^2.$$

Problem 3   Prove that the following inequalities hold for $x \geq 0$.

(1) $2x^3 - 3x^2 + 6x \geq 0$ (2) $2x(x^2 - 6) > 3(x^2 - 7)$

 Velocity

When point P moves along a straight line, let us define coordinates on this line and take it as the number line. If we take x as the coordinate of point P at time t, then x is a function of t. Let us take this function as

$$x = f(t).$$

The average velocity of point P from time t to $t + \Delta t$ is

$$\frac{\Delta x}{\Delta t} = \frac{f(t + \Delta t) - f(t)}{\Delta t}.$$

Let us designate the limit of this average velocity for $\Delta t \to 0$ as v, and then

$$v = \frac{dx}{dt} = \lim_{\Delta t \to 0} \frac{\Delta x}{\Delta t} = \lim_{\Delta t \to 0} \frac{f(t + \Delta t) - f(t)}{\Delta t}.$$

This v is called the **velocity** of point P at time t.

Demonstration 1   The coordinate of point P moving from the origin along the x-axis after t seconds can be expressed as

$$x = t^3 - 6t^2 + 9t$$

(1) Find the velocity of point P after $\frac{1}{3}$ second and 2 seconds.

(2) When does point P change its direction for the first time?

[Solution]   (1) Take v as the velocity of point P after t seconds, and then

$$v = \frac{dx}{dt} = 3t^2 - 12t + 9.$$

Therefore,   for $t = \frac{1}{3}$, $v = \frac{16}{3}$;

for $t = 2$, $v = -3$.

(2) $v = 3t^2 - 12t + 9 = 3(t - 1)(t - 3)$

t	0	$0 < t < 1$	1	$1 < t < 3$	3	$3 < t$
v		+	0	−	0	+

The direction of point P changes when the sign of v changes. Therefore, the direction changes for the first time after 1 second.

Problem 1   A train is moving along a straight track. The train goes $27t - 0.45t^2$ m in t seconds after the brakes are applied until it comes to a stop. How many seconds does it take the train to stop after the brakes are applied, and how many meters does it go?

Demonstration 2   The radius of a ball is expanding at a rate of 1 mm per second. Find the rate of change of the surface area and the volume of this ball when the radius is 4 cm. The radius at time $t = 0$ is 0.

[Solution]   Take r cm as the radius after t seconds, and then

$$r = 0.1t.$$

Take S cm^2 and V cm^3 as the surface area and the volume:

$$S = 4\pi r^2 = 0.04\pi t^2; \qquad\qquad V = \frac{4}{3}\pi r^3 = \frac{0.004}{3}\pi t^3.$$

Therefore, the rate of change of S and V with respect to t is

$$\frac{dS}{dt} = 0.08\pi t; \qquad\qquad \frac{dV}{dt} = 0.004\pi t^2.$$

Since $t = 40$ for $r = 4$, we obtain

$$(\frac{dS}{dt})_{t = 40} = 3.2\pi \text{ cm}^2/\text{s}; \qquad\qquad (\frac{dV}{dt})_{t = 40} = 6.4\pi \text{ cm}^3/\text{s}.$$

Problem 2   If we throw a stone at a still surface of a pond, circular ripples with the same center appear. At what rate is the area of the outermost circle increasing after three seconds, if the radius of that circle increases at the rate of 80 cm/second?

The rate of change of velocity with respect to time is called **acceleration**. Thus, if the velocity is expressed as

$$v = g(t),$$

the acceleration α is

$$\alpha = \frac{dv}{dt} = g'(t).$$

Example   The height of an object thrown upward at a velocity of v_0 m/sec after t seconds is $v_0 t - 4.9t^2$ m.

The velocity v m/sec after t seconds is $v = v_0 - 9.8t$.

The acceleration α m/sec^2 after t seconds is $\alpha = -9.8$.

Problem 3 The coordinates of two points P and Q moving along the x-axis from the origin after t seconds are $p = 2t^3 - 9t^2$ and $q = t^2 + 8t$, respectively. Take R as the midpoint of line segment PQ, and find the velocity and the acceleration of R after t seconds.

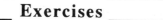

Exercises

1. Given the graph of the function $y = 2x^3 - 16x + 11$.

    (1) Find the equation of the tangent line at the point $(2, -5)$ on the graph.

    (2) Find the equation of the tangent line with a slope of -10.

2. The curve $y = ax^3 + bx^2 + cx + d$ is tangent to the line $y = x + 1$ at point $A(0, 1)$, and it is also tangent to the line $y = -2x + 10$. Find the values of the constants a, b, c, and d.

3. Determine when the following functions are increasing and decreasing, and find their extremes. Then draw the graphs.

    (1) $y = x(x - 3)^2$ (2) $y = 4 + 3x^2 - x^3$ (3) $y = -x^2(1 - x)^2$

4. The function $f(x) = x^3 + 3ax^2 + 3bx + c$ has a local maximum at $x = 1$ and a local minimum at $x = 2$.

    (1) Find the values of a and b.

    (2) Find the difference between the local maximum value and local minimum value.

5. Prove that the inequality $4x^3 - 3x^4 \leq 1$ holds for any real number x.

6. Find the minimum length of a line segment that connects the two curves $y = x^4 + 29$ and $y = 4x^3$ and lies parallel to the y-axis.

7. A piece of construction paper has the form of an equilateral triangle with a side of a. We want to make a box without a lid by cutting congruent quadrilaterals from the three corners and folding the remaining paper. What should the value of x be to make the volume of the box a maximum?

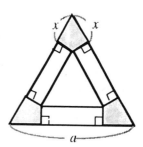

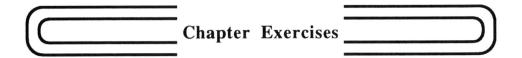

Chapter Exercises

A

1. The average rate of change of the function $y = ax^3 + 5x + 3$ from $x = 1$ to $x = 3$ is -8.

   (1) Find the value of the constant a.

   (2) Find the differential coefficient for $x = 3$.

2. Find the value of θ such that given the function $f(x) = x^2$, the following equality holds, provided that $h \neq 0$:

   $$f(x_1 + h) = f(x_1) + hf'(x_1 + \theta h)$$

3. Draw a tangent line to the curve $y = x^3$ at any point P on this curve and take Q and R as the points at which this tangent line intersects the x- and y-axes, respectively. Find $PQ : QR$.

4. Determine when the function $y = 6x^5 - 15x^4 + 10x^3 - 2$ is increasing and decreasing, and find its extremes. Then draw the graph.

5. Take A and B as the points at which the parabola $y = 9 - x^2$ intersects the x-axis. Consider a trapezoid $ABCD$ inscribed within the parabola and bounded by the line segment AB (with CD above AB). Find the coordinates of C and D such that the area of this trapezoid is a maximum.

6. A girl who is 1.6 m tall moves away from the base of a street lamp which is 4 m high at a velocity of 84 m/minute. Find the velocity of the top of her shadow. At what rate does the length of her shadow increase?

7. Given a square with sides of 15 cm. The length of each side is increasing by 2 cm/second.

   (1) Find the rate at which the area is increasing after 10 seconds.

   (2) Find the rate of increase at the moment when the area of this square becomes 400 cm^2.

B

1. Find the following limits for $f(x) = x^3$.

(1) $\lim_{h \to 0} \dfrac{f(a + 2h) - f(a)}{h}$

(2) $\lim_{h \to 0} \dfrac{f(a + h) - f(a - h)}{h}$

2. Two curves $y = x^3 + ax$ and $y = x^2 + bx + c$ pass through the point $(1, 2)$ and have a common tangent line at this point. Find the values of the constants a, b, and c.

3. Find the range of the real number k such that the function $f(x) = x^4 - 4x^3 + 2kx^2$ has no extreme.

4. Find the maximum and minimum values of the function $y = x^3 - 3x^2 + 2$ on the interval $0 \le x \le a$.

5. Find the maximum and minimum values of the function $y = 2x^3 - 3ax^2 + 1$ on the interval $0 \le x \le 3$, provided $a > 0$.

6. Given a regular tetrahedron $ABCD$ with edges of 12 cm. Take points P, Q, and R equidistant from point A on edges AB, AC, and AD, respectively. Draw perpendicular lines PP', QQ', and RR' from points P, Q and R to the base BCD. Find the length of AP if the volume of the triangular prism PQR-$P'Q'R'$ is a maximum.

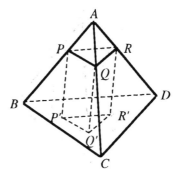

7. The equation $2x^3 - 3x^2 - 12x - a = 0$ has three different real solutions α, β, and γ, and $\alpha < \beta < \gamma$.

(1)   Find the range of the real number a.

(2)   Find the range of α, β, and γ.

CHAPTER 5

INTEGRATION AND ITS APPLICATIONS

SECTION 1.   INTEGRATION
SECTION 2.   APPLICATIONS OF DEFINITE INTEGRALS

It is said that integration originated with Archimedes (287?-212 B.C.). He computed the area of a figure bordered by a parabola and a straight line. The arguments on which the computation was based were so rigorous that seventeenth-century scholars referred to them as a "reductio ad absurdum."

The development of differentiation in the seventeenth century made it possible to regard integration as essentially the inverse of differentiation, and various integrations were easily computed. Later the term "calculus," which originally meant calculation, came to refer to differentiation and integration. The content of this chapter follows along the lines of these early ideas.

The principles of calculus were founded on processes of dynamics and extended to area and volume. They were not sufficiently rigorous to satisfy mathematicians. As a result, mathematicians went on to develop a new field, the theory of integration.

INTEGRATION

Indefinite Integrals

If the derivative of $F(x)$, a function in x, is $f(x)$, that is,

$$F'(x) = f(x)$$

$F(x)$ is called the **indefinite integral**, or **primitive**, of $f(x)$.

For example, the derivatives of the following functions in x

$$\frac{x^3}{3}, \quad \frac{x^3}{3} + 5, \quad \frac{x^3}{3} - 8$$

are all x^2, and therefore these functions are all indefinite integrals of x^2.

In general, if $F(x)$ is an indefinite integral of $f(x)$, then

$$(F(x) + C)' = F'(x) = f(x)$$

for any constant C. Therefore, $F(x) + C$ is also an indefinite integral of $f(x)$.

Conversely, all the indefinite integrals of $f(x)$ can be written in the form

$$F(x) + C$$

with an appropriate constant C. This is true because if we take $G(x)$ as an arbitrary indefinite integral of $f(x)$, then

$$F'(x) = f(x), \quad G'(x) = f(x).$$

Therefore,

$$\{G(x) - F(x)\}' = G'(x) - F'(x) = f(x) - f(x) = 0.$$

A function which has a derivative of 0 is a constant, so we can take C as that constant. Then,

$$G(x) - F(x) = C.$$

Thus,

$$G(x) = F(x) + C.$$

Accordingly, any indefinite integral of $f(x)$ can be expressed in the form

$$F(x) + C \quad \text{(provided that } C \text{ is a constant).}$$

An indefinite integral of $f(x)$ is generally designated by the symbol

$$\int f(x)dx \, *.$$

If we take $F(x)$ as any indefinite integral of $f(x)$, then

$$\int f(x)dx = F(x) + C.$$

C is called the **constant of integration**, and finding the indefinite integral is referred to as **integrating** $f(x)$.

(**Example**) $\displaystyle \int x^2 dx = \frac{x^3}{3} + C$ (C is a constant).

(**Problem 1**) Find the following indefinite integrals. Note that $\displaystyle \int dx$ means
$$\int 1dx.$$

$$(1) \quad \int 2x \, dx \qquad\qquad (2) \quad \int x^3 \, dx \qquad (3) \quad \int dx$$

If n is a positive integer or 0,

$$(x^{n+1})' = (n + 1)x^n.$$

Therefore,

$$\left(\frac{x^{n+1}}{n + 1} \right)' = x^n.$$

We can now present the following formula.

---

* The integral symbol $\displaystyle \int$ is read "integral."

Indefinite Integral of x^n

If n is a positive integer or 0, and we take C as a constant of integration, then

$$\int x^n \, dx = \frac{1}{n+1} x^{n+1} + C.$$

Demonstration 1   Find the following indefinite integrals:

(1) $\int x^7 \, dx$ (2) $\int x^9 \, dx$

[Solution]   (1) $\int x^7 \, dx = \frac{1}{7+1} x^{7+1} + C = \frac{1}{8} x^8 + C$

(2) $\int x^9 \, dx = \frac{1}{9+1} x^{9+1} + C = \frac{1}{10} x^{10} + C$

C is a constant of integration.

Note: Henceforth, we will omit the note about the constant of integration.

Problem 2   Find the following indefinite integrals:

(1) $\int x^4 \, dx$ (2) $\int x^6 \, dx$

We can derive formulas for indefinite integrals by using the differentiation formulas. If we take $F(x)$ and $G(x)$ as indefinite integrals of functions $f(x)$ and $g(x)$ and k as a constant, then we have

$$F'(x) = f(x); \qquad\qquad G'(x) = g(x);$$

$$\{kF(x)\}' = kF'(x) = kf(x);$$

$$\{F(x) + G(x)\}' = F'(x) + G'(x) = f(x) + g(x);$$

$$\{F(x) - G(x)\}' = F'(x) - G'(x) = f(x) - g(x).$$

From these equalities, we have the following formulas:

$$[\text{I}] \quad \int kf(x)\, dx = k \int f(x)\, dx$$

$$[\text{II}] \quad \int \{f(x) + g(x)\}\, dx = \int f(x)\, dx + \int g(x)\, dx$$

$$[\text{III}] \quad \int \{f(x) - g(x)\}\, dx = \int f(x)\, dx - \int g(x)\, dx$$

Problem 3   Prove that

$$\int \{f(x) + g(x) + h(x)\}\, dx = \int f(x)\, dx + \int g(x)\, dx + \int h(x)\, dx.$$

Demonstration 2   Find the indefinite integral $\int (9x^2 - 5x + 3)\, dx$.

[Solution]
$$\int (9x^2 - 5x + 3)\, dx = \int 9x^2\, dx - \int 5x\, dx + \int 3dx$$

$$= 9\int x^2\, dx - 5\int x\, dx + 3\int dx$$

$$= 9 \times \frac{x^3}{3} - 5 \times \frac{x^2}{2} + 3x + C$$

$$= 3x^3 - \frac{5}{2}x^2 + 3x + C$$

Problem 4   Find the following indefinite integrals:

(1) $\displaystyle\int (3x^2 - 2x - 1)\,dx$ (2) $\displaystyle\int (4x^3 - 6x + 5)\,dx$

(3) $\displaystyle\int (x - 1)(2x + 3)\,dx$ (4) $\displaystyle\int (t^4 - 2t^3 + t)\,dt$

(5) $\displaystyle\int (2y + 3)^2\,dy$

Demonstration 3   Find the indefinite integral of $x^2 - 4x$ that takes on the value -4 for $x = 3$.

[Solution]   If we take $F(x)$ as the function we want to find, then

$$F(x) = \int (x^2 - 4x)\,dx = \frac{x^3}{3} - 2x^2 + C.$$

From the condition $F(3) = -4$,

$$\frac{27}{3} - 2 \times 9 + C = -4.$$

Therefore,
$$C = 5.$$

Thus,
$$F(x) = \frac{x^3}{3} - 2x^2 + 5.$$

Problem 5   Find the indefinite integral of $2x + 7$ that takes on the value 10 for $x = 2$.

Problem 6   One of the indefinite integrals of $x^3 - ax$ is equal to 3 for $x = 0$ and equal to 1 for $x = 2$. Find the value of the constant a and the indefinite integral.

Definite Integrals

Integrals and Area

Indefinite integrals, which we have regarded as the inverse of differentiation, are related to the calculation of area. Let's examine how they are connected.

For example, in the figure to the right, let us consider the area of the figure bordered by the straight line of the graph of the function

$$f(x) = \frac{1}{2}x + 3$$

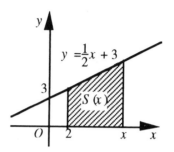

the x-axis, the line $x = 2$, and the line perpendicular to the x-axis through the point $(x, 0)$. If we take $S(x)$ as the area of this figure, we obtain

$$S(x) = \frac{1}{2}(4 + (\frac{1}{2}x + 3)) \times (x - 2) = \frac{1}{4}x^2 + 3x - 7.$$

Therefore,

$$S'(x) = \frac{1}{2}x + 3 \qquad \text{or} \qquad S'(x) = f(x).$$

In general, a function

$$y = f(x)$$

takes on positive values on the interval $a \leq x \leq b$ and the graph is continuous, as in the figure to the right. Take $S(x)$ as the area of the shaded region in the figure. The increment of $S(x)$ with respect to the increment of x,

$$\Delta S = S(x + \Delta x) - S(x)$$

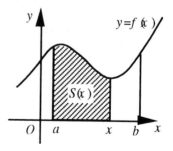

is the area of the figure on the interval from x to $x + \Delta x$ and between the curve $y = f(x)$ and the x-axis.

As you can see from the figure to the right, there is a value t between x and $x + \Delta x$ such that

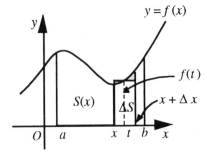

$$\Delta S = f(t)\,\Delta x$$

and as $\Delta x \to 0$, $t \to x$.

Therefore,

$$\lim_{\Delta x \to 0} \frac{\Delta S}{\Delta x} = \lim_{t \to x} f(t) = f(x). \qquad (1)$$

We examined the case for $\Delta x > 0$ above, but (1) holds even for $\Delta x < 0$. Thus,

$$\frac{dS}{dx} = S'(x) = f(x).$$

Therefore, if we take $F(x)$ as an indefinite integral of $f(x)$, then

$$S(x) = F(x) + C. \qquad (2)$$

Here, if we take $x = a$, then

$$S(a) = F(a) + C.$$

From the definition of $S(x)$, we know that $S(a) = 0$, so

$$C = -F(a). \qquad (3)$$

Substituting (3) into (2), we obtain

$$S(x) = F(x) - F(a).$$

If we let $x = b$, then we have $S(b)$, the area of the figure bordered by the curve $y = f(x)$, the x-axis, and the two lines $x = a$ and $x = b$.

Thus,

$$S(b) = F(b) - F(a).$$

Example   If $f(x) = x^2$, one indefinite
integral is

$$F(x) = \frac{x^3}{3}.$$

Therefore, the area of the
shaded region in the figure to
the right is

$$F(3) - F(1)$$

$$= \frac{3^3}{3} - \frac{1^3}{3} = \frac{26}{3}.$$

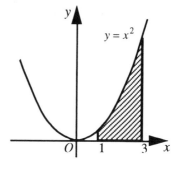

Problem 1   Find the following areas:

(1)   the area of the figure bordered by curve $y = x^3$, the x-axis, and
the two lines $x = 1$ and $x = 3$.

(2)   the area of the figure bordered by the curve $y = 6x - x^2$ and the
x-axis.

Definite Integrals

If $F(x)$ is an indefinite integral of $f(x)$, $F(b) - F(a)$ is called the **definite integral**
of $f(x)$ from a to b without regard to the value of a and b and is designated by

$$\int_a^b f(x)\, dx.$$

Here, a is referred to as the **lower limit** and b as the **upper limit**.

Definite Integral

If $F(x)$ is an indefinite integral of $f(x)$,

$$\int_a^b f(x)\, dx = F(b) - F(a).$$

The right side of the above formula, $F(b) - F(a)$, can be expressed as $[F(x)]_a^b$. Thus,

$$\int_a^b f(x)\,dx = [F(x)]_a^b = F(b) - F(a).$$

Demonstration   Find the following definite integrals:

(1) $\displaystyle\int_0^4 x^3\,dx$ (2) $\displaystyle\int_{-1}^3 (5x - x^2)\,dx$

[Solution]   (1) $\displaystyle\int_0^4 x^3\,dx = [\frac{x^4}{4}]_0^4 = \frac{4^4}{4} - 0 = 64$

(2) $\displaystyle\int_{-1}^3 (5x - x^2)\,dx = [\frac{5}{2}x^2 - \frac{x^3}{3}]_{-1}^3$

$$= (\frac{45}{2} - 9) - (\frac{5}{2} + \frac{1}{3}) = \frac{32}{3}$$

Problem 2   Find the following definite integrals:

(1) $\displaystyle\int_{-1}^2 x^4\,dx$ (2) $\displaystyle\int_{-2}^0 (4x^3 - 5)\,dx$

Problem 3   Find the following definite integrals:

(1) $\displaystyle\int_0^3 (t - 3)^2\,dt$ (2) $\displaystyle\int_3^{-1} (y^3 - 2y^2 + 4)\,dy$

Note: The value of a definite integral never changes, even if the variable x is replaced by other letters.

Topics for Enrichment: Approximate Sum of Definite Integrals

The function $y = f(x)$ is a continuous function that takes on positive values on the interval $a \leq x \leq b$. From our discussion on pages 145-146, we know that the value of a definite integral

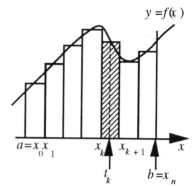

$$I = \int_a^b f(x)\, dx$$

is equal to the area S of the region bordered by the graph of $y = f(x)$, the x-axis, and the two lines $x = a$ and $x = b$.

Let's divide this interval into segments of the same length by selecting dividing points:

$$x_0 = a,\ x_1,\ x_2,\ \ldots,\ x_k,\ \ldots,\ x_n = b. \tag{1}$$

Thus,

$$x_k = a + k\frac{b - a}{n} \qquad (k = 0, 1, \ldots, n).$$

And let us take $t_0, t_1, \ldots, t_{n-1}$ as the midpoints of these segments. Thus,

$$t_k = \frac{x_k + x_{k+1}}{2} \qquad (k = 0, 1, \ldots, n - 1).$$

Now, $\qquad f(t_k)(x_{k+1} - x_k)$

is equal to the area of the shaded rectangle in the figure. Therefore, if we set

$$J = \sum_{k=0}^{n-1} f(t_k)\,(x_{k+1} - x_k), \tag{2}$$

J represents the area of the stairstep-shaped figure. As we allow n to become very large, or as we make the width of each segment very small, this area J approaches the area S. That is,

the value of J approaches the value

of the definite integral $\int_a^b f(x)\, dx.$

This also holds even if the division represented by (1) does not yield segments of equal width in (2), or if t_k is not the midpoint of x_k and x_{k+1}, as long as

$$x_k \leq t_k \leq x_{k+1} \qquad (k = 0, 1, ..., n - 1).$$

Moreover, we do not have to restrict $f(x)$ to positive values. As we make the segments narrow,

the value of J approaches the definite integral $\displaystyle\int_a^b f(x)\, dx.$

In this case, we refer to this value J as the **approximate sum** of $\displaystyle\int_a^b f(x)\, dx.$

Example $\quad f(x) = x^2$, $a = 0$, $b = 1$. Divide the interval $0 \leq x \leq 1$ by means of points $x_0, x_1, ..., x_n$, so that

$$x_k = \frac{k}{n} \qquad (k = 0, 1, ..., n).$$

Take x_k as t_k. Then

$$J = \sum_{k=0}^{n-1} f(t_k)\,(x_{k+1} - x_k) = \sum_{k=0}^{n-1} \left(\frac{k}{n}\right)^2 \cdot \frac{1}{n} = \frac{1}{n^3} \sum_{k=1}^{n-1} k^2.$$

From the result on page 80,

$$\sum_{k=1}^{n-1} k^2 = \frac{(n-1)\,n\,(2n-1)}{6}.$$

Therefore,

$$J = \frac{1}{n^3} \times \frac{(n-1)\,n\,(2n-1)}{6} = \frac{1}{6}\left(1 - \frac{1}{n}\right)\left(2 - \frac{1}{n}\right).$$

As we make n large and the segments narrow, the value of J approaches $\frac{1}{6} \times 1 \times 2 = \frac{1}{3}$, or

$$\int_0^1 f(x)\, dx = \int_0^1 x^2\, dx = \frac{1}{3}.$$

Properties of Definite Integrals

For definite integrals we can derive formulas similar to those for indefinite integrals.

$$[I] \quad \int_a^b kf(x)\, dx = k\int_a^b f(x)\, dx$$

$$[II] \quad \int_a^b \{f(x) + g(x)\}\, dx = \int_a^b f(x)\, dx + \int_a^b g(x)\, dx$$

$$[III] \quad \int_a^b \{f(x) - g(x)\}\, dx = \int_a^b f(x)\, dx - \int_a^b g(x)\, dx$$

Let's prove formula [II].

If we take $F(x)$ and $G(x)$ as indefinite integrals of $f(x)$ and $g(x)$, then one indefinite integral of $f(x) + g(x)$ is $F(x) + G(x)$. Therefore,

$$\int_a^b \{f(x) + g(x)\}\, dx = [F(x) + G(x)]_a^b$$

$$= \{F(b) + G(b)\} - \{F(a) + G(a)\}$$

$$= \{F(b) - F(a)\} + \{G(b) - G(a)\}$$

$$= \int_a^b f(x)\, dx + \int_a^b g(x)\, dx.$$

Problem 1 Prove formulas [I] and [III].

Example

$$\int_{-1}^{2} (x^2 + 6x - 2)\, dx = \int_{-1}^{2} x^2\, dx + 6\int_{-1}^{2} x\, dx - 2\int_{-1}^{2} dx$$

$$= [\frac{x^3}{3}]_{-1}^{2} + 6[\frac{x^2}{2}]_{-1}^{2} - 2[x]_{-1}^{2}$$

$$= 3 + 9 - 6 = 6$$

Problem 2   Find the following definite integrals:

(1) $\displaystyle\int_{-1}^{4} x^2(x - 2)\, dx$ (2) $\displaystyle\int_{-1}^{1} x(x - 2)^2\, dx$

We can also derive the following formulas for definite integrals.

[IV] $\displaystyle\int_{a}^{a} f(x)\, dx = 0$

[V] $\displaystyle\int_{a}^{b} f(x)\, dx = -\int_{b}^{a} f(x)\, dx$

[VI] $\displaystyle\int_{a}^{b} f(x)\, dx = \int_{a}^{c} f(x)\, dx + \int_{c}^{b} f(x)\, dx$

Problem 3   Prove formulas [IV]-[VI] above.

Demonstration 1   Find the definite integral $\displaystyle\int_0^3 |x(x-2)|\, dx$.

[**Solution**]   For $x \le 0$ and $2 \le x$,

$$|x(x-2)| = x(x-2)$$

$$= x^2 - 2x.$$

For $0 \le x \le 2$,

$$|x(x-2)| = -x(x-2)$$

$$= -x^2 + 2x.$$

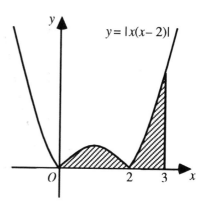

Therefore,

$$\int_0^3 |x(x-2)|\, dx = \int_0^2 (-x^2 + 2x)\, dx + \int_2^3 (x^2 - 2x)\, dx$$

$$= [-\frac{x^3}{3} + x^2]_0^2 + [\frac{x^3}{3} - x^2]_2^3$$

$$= (-\frac{8}{3} + 4) + (-\frac{8}{3} + 4)$$

$$= \frac{8}{3}.$$

Problem 4   Find the following definite integrals:

(1) $\displaystyle\int_2^5 |x-4|\, dx$ (2) $\displaystyle\int_0^3 |x^2 - 1|\, dx$

The following formula also holds.

[VII] $\displaystyle\frac{d}{dx}\int_a^x f(t)\, dt = f(x)$

[Proof]   If we take $F(t)$ as an indefinite integral of $f(t)$, then

$$\int_a^x f(t)\, dt = F(x) - F(a).$$

Since $F(a)$ is a constant, if we differentiate both sides with respect to x, we obtain

$$\frac{d}{dx} \int_a^x f(t)\, dt = F'(x) = f(x).$$

Demonstration 2   Find the function $f(x)$ and the constant a such that the following equality holds for an arbitrary x.

$$\int_a^x f(t)\, dt = 3x^2 + 2x - 5 \tag{1}$$

[Solution]   If we set $x = a$, then $\displaystyle\int_a^a f(t)\, dt = 0$. Therefore,

$$0 = 3a^2 + 2a - 5.$$

Therefore,
$$(3a + 5)(a - 1) = 0.$$

Thus,
$$a = -\frac{5}{3},\ 1.$$

We differentiate both sides of (1) with respect to x; then by using formula [VII] we obtain

$$f(x) = 6x + 2.$$

Problem 5   Find the value of $f(x)$ and the constant a if the equality

$$\int_1^x f(t)\, dt = x^3 + ax - 5$$

holds for any x.

Exercises

1. Find the following indefinite integrals:

   (1) $\displaystyle\int \frac{3}{4} dx$

   (2) $\displaystyle\int (6t - 3)\, dt$

   (3) $\displaystyle\int (6y^2 - 8y + 1)\, dy$

   (4) $\displaystyle\int x^2(3x^3 - 4)\, dx$

2. Find $f(x)$, if $f'(x) = 3x^2 - 4x + 5$ and $f(2) = 12$.

3. We were supposed to integrate $f(x)$ but we differentiated it by mistake, and we got the answer $3x^2 + 6x - 5$. Find the correct answer, if $f(0) = 3$.

4. The slope of a line tangent to the curve $y = f(x)$ at an arbitrary point $(x, f(x))$ on this curve and passing through the point $(0, 1)$ is $x^2 - 2x$. Find the function $f(x)$.

5. Find the following definite integrals:

   (1) $\displaystyle\int_2^1 x(x - 2)\, dx$

   (2) $\displaystyle\int_{-1}^1 (2x + 1)^3\, dx$

6. Find the definite integral $\displaystyle\int_{-2}^4 |x^2 - 2x - 3|\, dx$.

7. Find the value of the constant a such that the equality

   $$\int_0^a \{4x^3 - 2(3a + 2)x + 6\}\, dx = 0$$

   holds.

8. (1) Prove the following equations for any constant a.

   For an odd number n, $\displaystyle\int_{-a}^a x^n\, dx = 0$.

   For an even number n, $\displaystyle\int_{-a}^a x^n\, dx = 2\int_0^a x^n\, dx$.

   (2) Use the above results to find $\displaystyle\int_{-2}^2 x^3(x + 2)^2\, dx$.

2 APPLICATIONS OF DEFINITE INTEGRALS

  Area

Given the function $y = f(x)$, if $f(x) \geq 0$ on the interval $a \leq x \leq b$, the area S of the region bordered by the curve $y = f(x)$, the x-axis, and the two lines $x = a$ and $x = b$ is equal to the definite integral

$$\int_a^b f(x)\, dx.$$

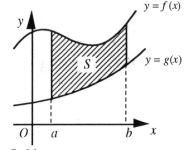

Next, if

$$g(x) \leq f(x)$$

on the interval $a \leq x \leq b$, let's consider how to find the area S of the figure bordered by the two curves $y = f(x)$ and $y = g(x)$ and the two lines $x = a$ and $x = b$.

If we assume that

$$0 \leq g(x) \leq f(x)$$

on the interval $a \leq x \leq b$, then the area S we want to find is

$$S = \int_a^b f(x)\, dx - \int_a^b g(x)\, dx = \int_a^b \{f(x) - g(x)\}\, dx.$$

On the interval $a \leq x \leq b$, if

$$g(x) \leq f(x)$$

and if $g(x)$ or $f(x)$ takes on a negative value, then translate the graphs of $y = f(x)$ and $y = g(x)$ upwards by m units along the positive direction of the y-axis so that both graphs are above the x-axis.

Since area is never changed by translation, the area S we want to find is

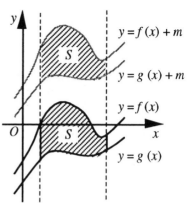

$$S = \int_a^b \{f(x) + m\} \, dx - \int_a^b \{g(x) + m\} \, dx$$

$$= \int_a^b \{f(x) - g(x)\} \, dx.$$

If $g(x) \leq f(x)$ on the interval $a \leq x \leq b$, the area S of the figure bordered by the two curves $y = f(x)$ and $y = g(x)$ and the two lines $x = a$ and $x = b$ is

$$S = \int_a^b \{f(x) - g(x)\} \, dx.$$

(**Demonstration 1**)   Find the area of the figure bordered by the two curves

$$y = 5 + 3x - x^2$$

and

$$y = x^2$$

and the two lines

$$x = 0 \quad \text{and} \quad x = 2.$$

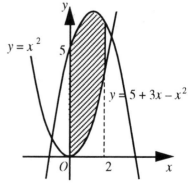

[Solution]   Set

$$f(x) = 5 + 3x - x^2, \quad g(x) = x^2.$$

Then $g(x) < f(x)$ on the interval $0 \le x \le 2$. Therefore, the area S we want to find is

$$S = \int_0^2 \{f(x) - g(x)\} \, dx = \int_0^2 (5 + 3x - 2x^2) \, dx$$

$$= [5x + \frac{3}{2}x^2 - \frac{2}{3}x^3]_0^2 = \frac{32}{3}.$$

Problem 1   Find the area of the figure bordered by the two parabolas $y = x^2 + 1$ and $y = -x^2 + 6x + 3$ and the two lines $x = 1$ and $x = 3$.

Demonstration 2   Find the area of the figure bordered by the parabola $y = x^2 - 2x - 3$ and the line $y = x + 1$.

[Solution]   In order to find the x coordinates of the points at which the parabola intersects the straight line, it is sufficient to solve the equation

$$x^2 - 2x - 3 = x + 1.$$

Solving it, we obtain

$$x = -1, 4.$$

The parabola lies below the straight line on the interval $-1 \le x \le 4$, so the area S we want to find is

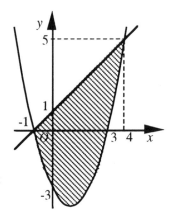

$$S = \int_{-1}^4 \{(x + 1) - (x^2 - 2x - 3)\} \, dx$$

$$= \int_{-1}^4 (-x^2 + 3x + 4) \, dx$$

$$= [-\frac{1}{3}x^3 + \frac{3}{2}x^2 + 4x]_{-1}^4 = \frac{125}{6}.$$

Problem 2   Find the area of the figure bordered by the two parabolas $y = x^2 + x$
and $y = 3 - x^2$.

If $f(x) = 0$ and $g(x) \le 0$ on the interval
$a \le x \le b$, the area S of the figure bordered by the curve
$y = g(x)$, the x-axis, and the two lines $x = a$ and $x = b$
is given by

$$S = \int_a^b \{0 - g(x)\}\, dx = -\int_a^b g(x)\, dx.$$

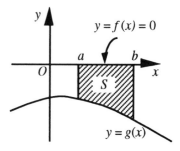

Problem 3   Find the area of the figure bordered by the parabola $y = x^2 - 3x - 4$,
the x-axis, the y-axis, and the line $x = 3$.

Problem 4   Do the following problems:

(1)   Show that $\displaystyle\int_\alpha^\beta (x - \alpha)(x - \beta)\, dx = -\frac{1}{6}(\beta - \alpha)^3$ holds.

(2)   If a quadratic equation $ax^2 + bx + c = 0$ has two distinct real
solutions α and β, show that the following equality holds:

$$\int_\alpha^\beta (ax^2 + bx + c)\, dx = -\frac{a}{6}(\beta - \alpha)^3.$$

(3)   Find the area of the figure bordered by the parabola $y = x^2 - 5x$
and the line $y = x + 2$.

Demonstration 3 Find the area of the figure bordered by the curve $y = x(x - 2)(x - 3)$ and the x-axis.

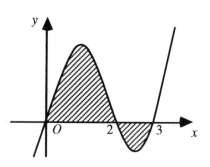

[Solution] The graph of this cubic curve is given to the right.

For $0 \le x \le 2$, $y \ge 0$;

for $2 \le x \le 3$, $y \le 0$.

Therefore, the area S we want to find is

$$S = \int_0^2 x(x - 2)(x - 3)\, dx - \int_2^3 x(x - 2)(x - 3)\, dx$$

$$= [\frac{1}{4}x^4 - \frac{5}{3}x^3 + 3x^2]_0^2 - [\frac{1}{4}x^4 - \frac{5}{3}x^3 + 3x^2]_2^3 = \frac{37}{12}.$$

Problem 5 Find the area of the figure bordered by the curve $y = 4x - x^3$ and the x-axis.

In the parabola $x^2 = ay$, if we interchange x and y, we obtain $y^2 = ax$. The graph of $y^2 = ax$ is created by reflecting the graph of $x^2 = ay$ with respect to the line $y = x$. It is a parabola in which the vertex is the origin and the axis is the x-axis.

Demonstration 4 Find the area of the figure bordered by the curve $y^2 = 3x$ and the line $y = 6$.

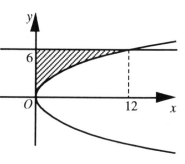

[Solution] The graph of $y^2 = 3x$ is a parabola which opens to the right, as in the figure at the right.

Therefore, the area S we want to find is

$$S = \int_0^6 x\, dy = \int_0^6 \frac{1}{3}y^2\, dy = [\frac{1}{9}y^3]_0^6 = 24.$$

Problem 6   Find the area of the figure bordered by the curve $x = y^2 + y - 2$ and the y-axis.

Problem 7   Find the area of the figure bordered by the curve $x = 6 - y^2$ and the line $y = x$.

 Volume

Integration can also be applied to finding the volume of solid bodies.

Take $S(x)$ as the area of a section cut from a solid body by plane X perpendicular to the x-axis at point x, as in the figure to the right.

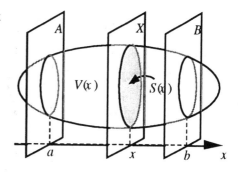

Let's consider the volume V of the part of a solid body between two planes A and B perpendicular to the x-axis.

Take a and b as the coordinates of the points at which planes A, B intersect the x-axis, and let $a < b$.

Take one point x on the interval $a \le x \le b$, and take $V(x)$ as the volume of the solid body between A and X. The increment of $V(x)$

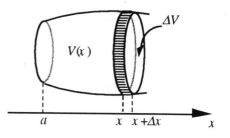

$$\Delta V = V(x + \Delta x) - V(x)$$

corresponding to Δx, the increment of x, is equal to the volume of the solid body between two planes perpendicular to the x-axis at two points x and $x + \Delta x$.

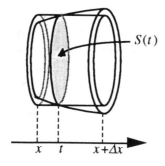 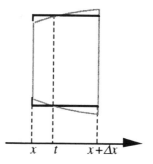

As you can see from the figure above, we can take a point t between x and $x + \Delta x$ such that

$$\Delta V = S(t)\, \Delta x$$

and

$$t \to x \text{ as } \Delta x \to 0.$$

Therefore,

$$V'(x) = \lim_{\Delta x \to 0} \frac{\Delta V}{\Delta x} = \lim_{t \to x} S(t) = S(x).$$

Thus, $V(x)$ is an indefinite integral of $S(x)$. Therefore,

$$\int_a^b S(x)\, dx = V(b) - V(a).$$

Since $V(a) = 0$,

$$V = V(b) = \int_a^b S(x)\, dx.$$

Demonstration 1 Find the volume of a pyramid in which the area of the base is S and the height is h.

[Solution] Take O as the vertex of the pyramid, and draw a perpendicular line OH from O to the base. Take an arbitrary plane X which intersects OH perpendicularly, and take x as the distance betweeen O and X. If we assume that $S(x)$ is the area of the section of the pyramid cut by X, S and $S(x)$ are the areas of similar figures with a ratio of $h : x$. Therefore,

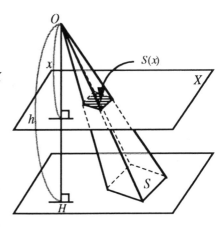

$$S(x) = S \times \frac{x^2}{h^2} = \frac{S}{h^2} x^2.$$

Thus, the volume of this pyramid is

$$V = \int_0^h S(x)\, dx = \int_0^h \frac{S}{h^2} x^2\, dx$$

$$= \frac{S}{h^2} [\frac{x^3}{3}]_0^h = \frac{1}{3} Sh.$$

(**Problem 1**) Find the volume of a cone in which the radius of the base is r and the height is h using the same method as in Demonstration 1.

(**Demonstration 2**) Find the volume of the solid body created by cutting a right cylinder by a plane that includes diameter AB of the base and that makes an angle of 45° with the base. The radius of the base is 10 cm.

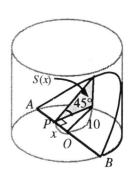

[**Solution**] Take O as the center of the base. Then take point P on diameter AB such that $OP = x$ cm. Let $S(x)$ cm^2 be the area of the section of this solid cut by a plane through point P and perpendicular to AB. This section is a right isosceles triangle in which the two sides that form the right angle are $\sqrt{10^2 - x^2}$ cm long. Therefore,

$$S(x) = \frac{1}{2}\sqrt{(10^2 - x^2)^2} = \frac{1}{2}(100 - x^2).$$

Thus, the volume V cm^3 of this solid is

$$V = 2\int_0^{10} S(x)\,dx = \int_0^{10} (100 - x^2)\,dx$$

$$= [100x - \frac{x^3}{3}]_0^{10} = \frac{2000}{3} \text{ cm}^3.$$

(**Problem 2**) Find the volume of a solid body created by cutting the right cylinder above by a plane that includes AB and makes an angle of 60° with the base.

Let's consider the solid of revolution created by revolving the curve

$$y = f(x)$$

on the interval $a \leq x \leq b$ about the x-axis.

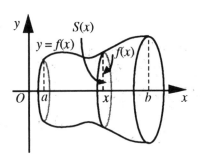

In this solid, the area $S(x)$ of a section in the figure to the right is

$$S(x) = \pi y^2 = \pi \{f(x)\}^2.$$

Therefore, the volume V can be expressed as:

The Volume of a Solid of Revolution

$$V = \pi \int_a^b y^2 \, dx = \pi \int_a^b \{f(x)\}^2 \, dx$$

(**Demonstration 3**) Find the volume of a ball of radius r.

[**Solution**] A ball of radius r can be considered as the solid of revolution created by revolving a semicircle

$$y = \sqrt{r^2 - x^2}$$

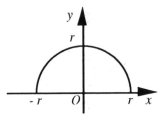

on the interval $-r \leq x \leq r$ about the x-axis.

Therefore, the volume V is

$$V = \pi \int_{-r}^r y^2 \, dx = \pi \int_{-r}^r (r^2 - x^2) \, dx = 2\pi \int_0^r (r^2 - x^2) \, dx$$

$$= 2\pi \, [r^2 x - \frac{x^3}{3}]_0^r = \frac{4}{3} \pi r^3.$$

Problem 3 Find the volumes of the following solid bodies using the formula for the volume of a solid of revolution:

(1) a right cone in which the radius of the base is r and the height is h.

(2) a right circular truncated cone in which the radius of the upper base is a, the radius of the lower base is b, and the height is h.

The volume V of a solid of revolution created by revolving a figure bordered by the curve

$$x = g(y)$$

the y-axis, and the two lines $y = c$ and $y = d$ about the y-axis can be thought of as

$$V = \pi \int_c^d x^2\, dy = \pi \int_c^d \{g(y)\}^2\, dy$$

by interchanging the roles of x and y.

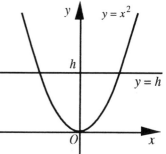

Example The volume V of a solid of revolution created by revolving the figure bordered by the parabola $y = x^2$ and the line $y = h$ about the y-axis is

$$V = \pi \int_0^h x^2\, dy$$

$$= \pi \int_0^h y\, dy = \frac{\pi h^2}{2}.$$

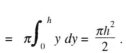

Reference: The Solid of Revolution of an Ellipse

As you learned in Algebra and Geometry, the curve represented by an equation of the form

$$\frac{x^2}{a^2} + \frac{y^2}{b^2} = 1$$

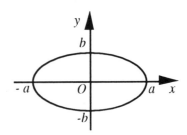

where a and b are positive integers, is called an ellipse, and it has the form illustrated at the right.

Let's find the volume V of the solid of revolution created by revolving this curve about the y-axis. From

$$\frac{x^2}{a^2} + \frac{y^2}{b^2} = 1,$$

we get

$$x^2 = a^2(1 - \frac{y^2}{b^2}).$$

Therefore,

$$V = \pi\int_{-b}^{b} x^2 \, dy = 2\pi\int_{0}^{b} a^2(1 - \frac{y^2}{b^2}) \, dy$$

$$= 2\pi a^2[y - \frac{y^3}{3b^2}]_0^b = \frac{4}{3}\pi a^2 b.$$

Problem   Find the volume of the solid of revolution created by revolving the above ellipse around the x-axis.

Demonstration 4   Find the volume of the solid of revolution created by revolving the figure bordered by the parabola $y = 4x - x^2$ and the straight line $y = x$ about the x-axis.

[Solution]   The x-coordinates of the points at which the parabola intersects the straight line are given by

$$4x - x^2 = x, \quad x^2 - 3x = 0$$

as

$$x = 0, 3.$$

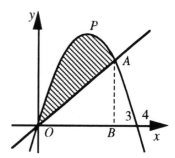

Take V_1 and V_2 as the volumes of the solids of revolution created by revolving figure $OPAB$ and triangle OAB about the x-axis. The volume V we want to find is

$$V = V_1 - V_2 \;=\; \pi \int_0^3 (4x - x^2)^2\, dx - \pi \int_0^3 x^2\, dx$$

$$=\; \pi \int_0^3 (15x^2 - 8x^3 + x^4)\, dx$$

$$=\; \pi [5x^3 - 2x^4 + \tfrac{1}{5}x^5]_0^3 \;=\; \frac{108}{5}\, \pi.$$

Example 4   Find the volume of the solid of revolution created by revolving the figure bordered by the two parabolas $y = x^2$ and $y^2 = x$ about the x-axis.

Example 5   Find the volume of the solid of revolution created by revolving the figure bordered by the two parabolas $y = x^2$ and $y = \dfrac{3}{4}x^2 + \dfrac{1}{2}$ about the y-axis.

Example 6   Sector OAB, with a radius of 6 and a central angle of 60°, is located as shown in the figure to the right. Find the solids of revolution created by revolving this figure about:

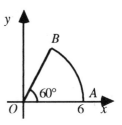

(1)   the x-axis   (2) the y-axis

   Velocity and Distance

An object is moving along a straight line. Take $s(t)$ as the coordinate of this point on the line at time t, and take $v(t)$ as the velocity. Then

$$v(t) = s'(t).$$

Therefore,

$$s(t) = \int v(t)\, dt.$$

Thus, for times t_1 and t_2,

$$s(t_2) = s(t_1) + \int_{t_1}^{t_2} v(t)\, dt.$$

This formula shows the relation between the location and the velocity at times t_1 and t_2. From this formula, we know the change in the location of a moving object from time t_1 to t_2.

Demonstration 1   A train moves along a straight track at a velocity of 27 m/sec. From the moment the brakes are applied to the moment the train comes to a stop, the velocity $v(t)$ m/sec at t seconds after the brakes were applied can be expressed as

$$v(t) = 27 - t \quad (0 \le t \le 27).$$

How many seconds will it take this train to stop after the brakes are applied? How many meters will it cover before stopping?

[Solution] Take $s(t)$ m as the distance the train travels after the brakes are applied, and then

$$v(t) = 27 - t.$$

Therefore,

$$s(t) = \int (27 - t)\, dt.$$

The train will stop when

$$v(t) = 0.$$

Thus,

$$t = 27.$$

Therefore, the train will stop after 27 seconds.

The distance the train will travel in that time is

$$s(27) \;=\; s(0) + \int_0^{27} (27 - t)\, dt$$

$$=\; [27t - \frac{1}{2} t^2]_0^{27} \;=\; 364.5 \;.$$

Thus, it will go 364.5 m.

(Problem 1) If we throw a ball along a straight line at an initial velocity of 4 m/second, the velocity of the ball decreases every second by 0.5 m/second. How far does the ball go?

(Demonstration 2) The velocity v of point P moving along the x-axis can be expressed as

$$v = 4 - 2t.$$

The coordinate of P at $t = 0$ is $x = 1$. Find the location of P after t seconds, and the distance that P actually covers from $t = 0$ to $t = 3$.

[**Solution**]   The coordinate $x(t)$ of P after t seconds is

$$x(t) \;=\; 1 + \int_0^t (4 - 2t)\, dt$$

$$=\; 1 + 4t - t^2.$$

The actual distance covered from $t = 0$ to $t = 3$ is

$$\int_0^3 |4 - 2t|\, dt \;=\; \int_0^2 (4 - 2t)\, dt + \int_2^3 (2t - 4)\, dt$$

$$=\; [4t - t^2]_0^2 + [t^2 - 4t]_2^3$$

$$=\; 5.$$

Problem 2   The velocity $v(t)$ m/sec of an object thrown upward at an initial velocity of 30 m/sec from a point 8 m above the ground can be expressed by the following formula.

$$v(t) = 30 - 10t$$

(1) Find the object's height above the ground after 1 second and after 4 seconds.

(2) Find the distance this object covers in the first 5 seconds.

Exercises

1. Find the areas of the figures bordered by the x-axis and the following curves:

(1) $y = x^2 - 2x - 3$ (2) $y = 3x^2 - x^3$

(3) $y = x^4 - 2x^2 + 1$

2. Sketch the region expressed by the simultaneous inequalities $y \le 3x + 6$, $x + y + 2 \ge 0$, and $y \le 22 - 3x - x^2$, and find its area.

3. We want to bisect the figure bounded by the two parabolas $y = x^2 - x + 1$ and $y = -2x^2 + 5x + 1$ with a straight line perpendicular to the x-axis. Find the equation of that line.

4. Take A and B as the tangent points of two tangents to the parabola $y = x^2 - 2x + 4$ which originate at the origin. Find the area of the region bordered by tangent lines OA and OB and the parabola.

5. Find the volume of the solid of revolution created by revolving the figure bordered by the graph of the function $f(x) = x^2(1 - x)$ and the x-axis around the x-axis.

6. Find the value of constant m such that the volumes of the solids of revolution created by revolving the figure bordered by the curve $y = \sqrt{x}$ and the line $y = mx$ about the x-axis and the y-axis are equal.

7. The relation between the velocity v and the time t for point P moving along a straight line is given as in the graph to the right.

    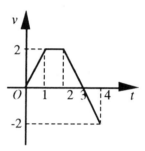

    (1) Find a formula to express the relation between the distance x of P from the origin and the time t. If $t = 0$, $x = 2$.

    (2) Find the actual distance covered by P from $t = 0$ to $t = 4$.

Chapter Exercises

A

1. Find the function $f(x)$ such that $f'(x) = x^2 - 4x$ and $f(1) = -1$.

2. Find functions $f(x)$ and $g(x)$ satisfying $\dfrac{d}{dx}\{f(x) + g(x)\} = 3$, $\dfrac{d}{dx}\{f(x)g(x)\} = 4x + 1$, $f(0) = -3$, and $g(0) = 2$.

3. Express the definite integral $\displaystyle\int_{-1}^{1} (x^3 - ax - b)^2\, dx$ in terms of a and b, and find the values of real numbers a and b such that the value of this integral is a minimum.

4. Find the area of the figure bordered by the curve $x = y^2 - 4$ and the y-axis.

5. Define constants a and b such that the curve $y = x^3 - x^2 - ax + b$ is tangent to the straight line $x - 2y + 1 = 0$ at the point $(1, 1)$. Then find the area of the figure bordered by this curve and straight line.

6. Given a semicircle with diameter AB. Take an arbitrary point P on AB. Consider chord PQ, perpendicular to AB and passing through P. Then erect a right isosceles triangle PQR perpendicular to the semicircle such that $PQ = PR$. Find the volume of the solid body that $\triangle PQR$ passes through when point P moves from A to B, given that $AB = 6$.

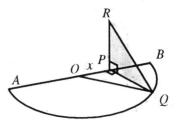

7. Find the volume of the solid body created by revolving the figure bordered by the parabola $y = 3 - x^2$ and the line $y = 2$ about the x-axis.

8. The figure to the right is a semicircle with a radius of 1, and the angles formed by chord AP with diameter AB and by chord AQ with diameter AB are 30° and 60°, respectively. Find the volume of the solid of revolution created by revolving the figure bordered by the chords AP, AQ, and arc PQ about diameter AB.

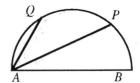

B

1. Prove that the inequality $\left(\int_0^1 f(x)\, dx\right)^2 < \int_0^1 \{f(x)\}^2\, dx$ holds if $f(x)$ is a linear function.

2. Find the extremes of the function $F(x) = \int_0^x t(t^2 - 1)\, dt$.

3. Find the function $f(x)$ which satisfies the following equality:

$$f(x) = x + \int_0^2 f(t)\, dt.$$

4. Find the value of a such that the areas of the two regions bordered by the curve $y = x(x-2)(x-a)$ and the x-axis are equal, provided that $a > 0$.

5. Find the volume of the solid of revolution created by revolving the figure bordered by the parabola $y = 4 - x^2$ and the x-axis about the line $y = -1$. Then find the volume if this figure is revolved about the line $y = 5$.

6. Find the volume of the solid of revolution created by revolving the figure bordered by the curve $y = x^2 - x - 2$ and the straight line $y = x + 1$ about the x-axis.

7. A semi-spherical container of radius r is full of water. If this container is tilted at an angle of 30°, find the quantity of water poured out of the container.

8. When a container with a height of 12 cm is cut by a plane at a distance of h cm from the base, the section is a square with sides of $\sqrt{h}$ cm. If water runs into this container at 8 cm^3/second, how many seconds will it take to fill the container?

9. Points P and Q move along the x-axis, starting simultaneously from the origin. The velocity u of P and the velocity v of Q after t seconds are

$$u = 2t, \quad v = 3t(4 - t).$$

When will P and Q meet for the first time? Find the total distance covered by P and Q by that time.

Answers to Chapter Exercises

Chapter 1. Exponential and Logarithmic Functions 25-27

A

1. (1) 64   (2) $\frac{1}{4}$
   (3) 2   (4) 128
2. (1) $a^{\frac{7}{6}}b$ (2) $x^{-\frac{1}{2}}y^{\frac{1}{2}}$
3. (1) 14   (2) 194
4. $y = x^{\frac{4}{3}}$
5. (1) $-\frac{1}{2}$ (2) -5   (3) 2
6. (1) $3m + \frac{1}{2}n$ (2) $\frac{1}{2}m - 2n$
   (3) $2 + m + n$ (4) $m + 3n$
7. (1) 4   (2) $\frac{1}{27}$ (3) ±100

B

1. (1) $a - \frac{1}{a}$ (2) 1
2. (1) $\frac{16}{3}$ (2) $\frac{7}{3}$
3. 3
4. 1
6. (1) 11th decimal place
   (2) 2nd decimal place
7. (1) 12   (2) 8
8. (1) $3 < x \le 30$ (2) $x \ge \frac{1}{8}$
9. $100, \frac{1}{10}$

Chapter 2.   Trigonometric Functions 61-63

A

1. (1) $-\frac{\sqrt{3}}{2}$ (2) 1   (3) $\frac{\sqrt{3}}{2}$
2. (1) 3rd quadrant   (2) 4th quadrant
3. (1) $-k$ (2) $-\sqrt{1 - k^2}$
4. (1) $-\frac{7}{2}$ (2) $\frac{3}{2}$
5. (1) $-\frac{4 + 6\sqrt{6}}{25}$ (2) $\frac{-7 + 48\sqrt{6}}{125}$
6. (1) $\frac{4}{3}\pi, \frac{5}{3}\pi$ (2) $\frac{\pi}{2}, \frac{3}{2}\pi$

B

3.
$$\begin{cases} \sin\theta = \dfrac{\sqrt{6} - \sqrt{2}}{4} \\ \cos\theta = \dfrac{\sqrt{6} + \sqrt{2}}{4} \end{cases}$$

$$\begin{cases} \sin\theta = \dfrac{\sqrt{2} - \sqrt{6}}{4} \\ \cos\theta = -\dfrac{\sqrt{2} + \sqrt{6}}{4} \end{cases}$$

4. (1) 3π (2) $\frac{\pi}{2}$
5. (1) maximum: 2
      minimum: -2
   (2) maximum: $\sqrt{7}$
      minimum: $-\sqrt{7}$
7. $\sin\theta = 0, \frac{1}{2}$
   $\theta = 0, \frac{\pi}{6}, \frac{5}{6}\pi, \pi$
8. (1) $\frac{\sqrt{6}}{2}, \frac{\sqrt{2}}{2}$
   (2) $\sin 15° = \dfrac{\sqrt{6} - \sqrt{2}}{4}$
      $\sin 75° = \dfrac{\sqrt{6} + \sqrt{2}}{4}$

Chapter 3.   Progressions

A

1. -2
2. $a = 10, b = 25, c = 40$ or $a = -6, b = 9,$
    $c = 24$
3. (1) $\frac{n}{2}(2n^2 + 7n - 3)$

    (2) $\frac{20}{81}(10^n - 1) - \frac{2}{9}n$

4. (1) $\frac{n(n + 1)(n + 2)}{6}$ (2) $\frac{2n}{n + 1}$

5. (1) n^2
6. (1) Geometric progression with a 1st
    term of 1 and a common ratio of
    $-\frac{1}{2}$

    (2) $b_n = (-\frac{1}{2})^{n-1}$

    $a_n = \frac{2}{3}\{1 - (-\frac{1}{2})^{n-1}\}$

B

1. (1) $2n \cdot 3^n$

    (2) $\frac{(2n - 1)3^{n+1} + 3}{2}$

2. The first term: $\frac{n^2 - n + 2}{2}$

    Sum: $\frac{n(n^2 + 1)}{2}$

    $a_n = 3(n + 1)$

4. (1) $b_{n+1} = 2b_n + 1$

    (2) $a_n = \frac{1}{2n - 1}$

5. (1) $a_{n+1} = a_n + 2n + 3$

    (2) $a_n = n^2 + 2n + 1$

Chapter 4.   Differentiation and Its Applications

A

1. (1) -1   (2) -22
2. $\frac{1}{2}$
3. $PQ : QR = 1 : 2$
5. C (1,8), D (-1, 8)
6. Velocity of shadow: 140 m/min.
    Rate of increase of length of
    shadow: 56 m/min.
7. (1) $140 \text{ cm}^2/\text{s}$ (2) $80 \text{ cm}^2/\text{s}$

B

1. (1) $6a^2$ (2) $6a^2$
2. $a = 1, b = 2, c = -1$
3. $k \geq \frac{9}{4}, k = 0$

4. For $0 < a \leq 2,$
    maximum: 2,
    minimum: $a^3 - 3a^2 + 2;$
    for $2 < a < 3,$
    maximum: 2
    minimum: -2;
    for $3 \leq a,$
    maximum: $a^3 - 3a^2 + 2$
    minimum: -2.
5. For $3 \leq a,$
    maximum: 1
    minimum: $55 - 27a;$
    for $2 < a < 3,$
    maximum: 1
    minimum: $1 - a^3;$
    for $0 < a \leq 2,$
    maximum: $55 - 27a$
    minimum: $1 - a^3.$
6. AP = 8cm

7. (1) $-20 < a < 7$ (2) $\begin{cases} -\frac{5}{2} < \alpha < -1 \\ -1 < \beta < 2 \\ 2 < \gamma < \frac{7}{2} \end{cases}$

Chapter 5. Integration and Its Applications

A

1. $f(x) = \dfrac{x^3}{3} - 2x^2 + \dfrac{2}{3}$

2. $f(x) = 2x - 3, \; g(x) = x + 2$

3. $a = \dfrac{3}{5}, \; b = 0$

4. $\dfrac{32}{3}$

5. $a = \dfrac{1}{2}, \; b = \dfrac{3}{2}, \;$ area: $\dfrac{4}{3}$

6. 18

7. $\dfrac{32}{5}\pi$

8. $\dfrac{2\pi}{3}$

B

2. Local minimum: $F(-1) = F(1) = -\dfrac{1}{4}$
   local maximum: $F(0) = 0$

3. $f(x) = x - 2$

4. $a = 4, \; 1$

5. For $y = -1, \dfrac{832}{15}\pi$;

   for $y = 5, \dfrac{1088}{15}\pi$.

6. $\dfrac{611}{30}\pi$

7. $\dfrac{11}{24}\pi r^3$

8. After 9 seconds

9. After 5 seconds

   P: 25

   Q: 39

Index

Greek Letters

Capital	Small		Capital	Small	
A	α	alpha	N	ν	nu
B	β	beta	Ξ	ξ	xi
Γ	γ	gamma	O	o	omicron
Δ	δ	delta	Π	π	pi
E	ε	epsilon	P	ρ	rho
Z	ζ	zeta	Σ	σ, ς	sigma
H	η	eta	T	τ	tau
Θ	θ, ϑ	theta	Y	υ	upsilon
I	ι	iota	Φ	φ, ϕ	phi
K	κ	kappa	X	χ	chi
Λ	λ	lambda	Ψ	ψ	psi
M	μ	mu	W	ω	omega

Table of Squares, Square Roots, and Reciprocals

n	n^2	$\sqrt{n}$	$\sqrt{10n}$	$\frac{1}{n}$	n	n^2	$\sqrt{n}$	$\sqrt{10n}$	$\frac{1}{n}$
1	1	1.0000	3.1623	1.0000	51	2601	7.1414	22.5832	0.0196
2	4	1.4142	4.4721	0.5000	52	2704	7.2111	22.8035	0.0192
3	9	1.7321	5.4772	0.3333	53	2809	7.2801	23.0217	0.0189
4	16	2.0000	6.3246	0.2500	54	2916	7.3485	23.2379	0.0185
5	25	2.2361	7.0711	0.2000	55	3025	7.4162	23.4521	0.0182
6	36	2.4495	7.7460	0.1667	56	3136	7.4833	23.6643	0.0179
7	49	2.6458	8.3666	0.1429	57	3249	7.5498	23.8747	0.0175
8	64	2.8284	8.9443	0.1250	58	3364	7.6158	24.0832	0.0172
9	81	3.0000	9.4868	0.1111	59	3481	7.6811	24.2899	0.0169
10	100	3.1623	10.0000	0.1000	60	3600	7.7460	24.4949	0.0167
11	121	3.3166	10.4881	0.0909	61	3721	7.8102	24.6982	0.0164
12	144	3.4641	10.9545	0.0833	62	3844	7.8740	24.8998	0.0161
13	169	3.6056	11.4018	0.0769	63	3969	7.9373	25.0998	0.0159
14	196	3.7417	11.8322	0.0714	64	4096	8.0000	25.2982	0.0156
15	225	3.8730	12.2474	0.0667	65	4225	8.0623	25.4951	0.0154
16	256	4.0000	12.6491	0.0625	66	4356	8.1240	25.6905	0.0152
17	289	4.1231	13.0384	0.0588	67	4489	8.1854	25.8844	0.0149
18	324	4.2426	13.4164	0.0556	68	4624	8.2462	26.0768	0.0147
19	361	4.3589	13.7840	0.0526	69	4761	8.3066	26.2679	0.0145
20	400	4.4721	14.1421	0.0500	70	4900	8.3666	26.4575	0.0143
21	441	4.5826	14.4914	0.0476	71	5041	8.4261	26.6458	0.0141
22	484	4.6904	14.8324	0.0455	72	5184	8.4853	26.8328	0.0139
23	529	4.7958	15.1658	0.0435	73	5329	8.5440	27.0185	0.0137
24	576	4.8990	15.4919	0.0417	74	5476	8.6023	27.2029	0.0135
25	625	5.0000	15.8114	0.0400	75	5625	8.6603	27.3861	0.0133
26	676	5.0990	16.1245	0.0385	76	5776	8.7178	27.5681	0.0132
27	729	5.1962	16.4317	0.0370	77	5929	8.7750	27.7489	0.0130
28	784	5.2915	16.7332	0.0357	78	6084	8.8318	27.9285	0.0128
29	841	5.3852	17.0294	0.0345	79	6241	8.8882	28.1069	0.0127
30	900	5.4772	17.3205	0.0333	80	6400	8.9443	28.2843	0.0125
31	961	5.5678	17.6068	0.0323	81	6561	9.0000	28.4605	0.0123
32	1024	5.6569	17.8885	0.0313	82	6724	9.0554	28.6356	0.0122
33	1089	5.7446	18.1659	0.0303	83	6889	9.1104	28.8097	0.0120
34	1156	5.8310	18.4391	0.0294	84	7056	9.1652	28.9828	0.0119
35	1225	5.9161	18.7083	0.0286	85	7225	9.2195	29.1548	0.0118
36	1296	6.0000	18.9737	0.0278	86	7396	9.2736	29.3258	0.0116
37	1369	6.0828	19.2354	0.0270	87	7569	9.3274	29.4958	0.0115
38	1444	6.1644	19.4936	0.0263	88	7744	9.3808	29.6648	0.0114
39	1521	6.2450	19.7484	0.0256	89	7921	9.4340	29.8329	0.0112
40	1600	6.3246	20.0000	0.0250	90	8100	9.4868	30.0000	0.0111
41	1681	6.4031	20.2485	0.0244	91	8281	9.5394	30.1662	0.0110
42	1764	6.4807	20.4939	0.0238	92	8464	9.5917	30.3315	0.0109
43	1849	6.5574	20.7364	0.0233	93	8649	9.6437	30.4959	0.0108
44	1936	6.6332	20.9762	0.0227	94	8836	9.6954	30.6594	0.0106
45	2025	6.7082	21.2132	0.0222	95	9025	9.7468	30.8221	0.0105
46	2116	6.7823	21.4476	0.0217	96	9216	9.7980	30.9839	0.0104
47	2209	6.8557	21.6795	0.0213	97	9409	9.8489	31.1448	0.0103
48	2304	6.9282	21.9089	0.0208	98	9604	9.8995	31.3050	0.0102
49	2401	7.0000	22.1359	0.0204	99	9801	9.9499	31.4643	0.0101
50	2500	7.0711	22.3607	0.0200	100	10000	10.0000	31.6228	0.0100

Table of Common Logarithms (1)

	0	1	2	3	4	5	6	7	8	9	1	2	3	4	5	6	7	8	9
1.0	.0000	.0043	.0086	.0128	.0170	.0212	.0253	.0294	.0334	.0374	4	8	12	17	21	25	29	33	37
1.1	.0414	.0453	.0492	.0531	.0569	.0607	.0645	.0682	.0719	.0755	4	8	11	15	19	23	26	30	34
1.2	.0792	.0828	.0864	.0899	.0934	.0969	.1004	.1038	.1072	.1106	3	7	10	14	17	21	24	28	31
1.3	.1139	.1173	.1206	.1239	.1271	.1303	.1335	.1367	.1399	.1430	3	6	10	13	16	19	23	26	29
1.4	.1461	.1492	.1523	.1553	.1584	.1614	.1644	.1673	.1703	.1732	3	6	9	12	15	18	21	24	27
1.5	.1761	.1790	.1818	.1847	.1875	.1903	.1931	.1959	.1987	.2014	3	6	8	11	14	17	20	22	25
1.6	.2041	.2068	.2095	.2122	.2148	.2175	.2201	.2227	.2253	.2279	3	5	8	11	13	16	18	21	24
1.7	.2304	.2330	.2355	.2380	.2405	.2430	.2455	.2480	.2504	.2529	2	5	7	10	12	15	17	20	22
1.8	.2553	.2577	.2601	.2625	.2648	.2672	.2695	.2718	.2742	.2765	2	5	7	9	12	14	16	19	21
1.9	.2788	.2810	.2833	.2856	.2878	.2900	.2923	.2945	.2967	.2989	2	4	7	9	11	13	16	18	20
2.0	.3010	.3032	.3054	.3075	.3096	.3118	.3139	.3160	.3181	.3201	2	4	6	8	11	13	15	17	19
2.1	.3222	.3243	.3263	.3284	.3304	.3324	.3345	.3365	.3385	.3404	2	4	6	8	10	12	14	16	18
2.2	.3424	.3444	.3464	.3483	.3502	.3522	.3541	.3560	.3579	.3598	2	4	6	8	10	12	14	15	17
2.3	.3617	.3636	.3655	.3674	.3692	.3711	.3729	.3747	.3766	.3784	2	4	6	7	9	11	13	15	17
2.4	.3802	.3820	.3838	.3856	.3874	.3892	.3909	.3927	.3945	.3962	2	4	5	7	9	11	12	14	16
2.5	.3979	.3997	.4014	.4031	.4048	.4065	.4082	.4099	.4116	.4133	2	3	5	7	9	10	12	14	15
2.6	.4150	.4166	.4183	.4200	.4216	.4232	.4249	.4265	.4281	.4298	2	3	5	7	8	10	11	13	15
2.7	.4314	.4330	.4346	.4362	.4378	.4393	.4409	.4425	.4440	.4456	2	3	5	6	8	9	11	13	14
2.8	.4472	.4487	.4502	.4518	.4533	.4548	.4564	.4579	.4594	.4609	2	3	5	6	8	9	11	12	14
2.9	.4624	.4639	.4654	.4669	.4683	.4698	.4713	.4728	.4742	.4757	1	3	4	6	7	9	10	12	13
3.0	.4771	.4786	.4800	.4814	.4829	.4843	.4857	.4871	.4886	.4900	1	3	4	6	7	9	10	11	13
3.1	.4914	.4928	.4942	.4955	.4969	.4983	.4997	.5011	.5024	.5038	1	3	4	6	7	8	10	11	12
3.2	.5051	.5065	.5079	.5092	.5105	.5119	.5132	.5145	.5159	.5172	1	3	4	5	7	8	9	11	12
3.3	.5185	.5198	.5211	.5224	.5237	.5250	.5263	.5276	.5289	.5302	1	3	4	5	6	8	9	10	12
3.4	.5315	.5328	.5340	.5353	.5366	.5378	.5391	.5403	.5416	.5428	1	3	4	5	6	8	9	10	11
3.5	.5441	.5453	.5465	.5478	.5490	.5502	.5514	.5527	.5539	.5551	1	2	4	5	6	7	9	10	11
3.6	.5563	.5575	.5587	.5599	.5611	.5623	.5635	.5647	.5658	.5670	1	2	4	5	6	7	8	10	11
3.7	.5682	.5694	.5705	.5717	.5729	.5740	.5752	.5763	.5775	.5786	1	2	3	5	6	7	8	9	10
3.8	.5798	.5809	.5821	.5832	.5843	.5855	.5866	.5877	.5888	.5899	1	2	3	5	6	7	8	9	10
3.9	.5911	.5922	.5933	.5944	.5955	.5966	.5977	.5988	.5999	.6010	1	2	3	4	5	7	8	9	10
4.0	.6021	.6031	.6042	.6053	.6064	.6075	.6085	.6096	.6107	.6117	1	2	3	4	5	7	8	9	10
4.1	.6128	.6138	.6149	.6160	.6170	.6180	.6191	.6201	.6212	.6222	1	2	3	4	5	6	7	8	9
4.2	.6232	.6243	.6253	.6263	.6274	.6284	.6294	.6304	.6314	.6325	1	2	3	4	5	6	7	8	9
4.3	.6335	.6345	.6355	.6365	.6375	.6385	.6395	.6405	.6415	.6425	1	2	3	4	5	6	7	8	9
4.4	.6435	.6444	.6454	.6464	.6474	.6484	.6493	.6503	.6513	.6522	1	2	3	4	5	6	7	8	9
4.5	.6532	.6542	.6551	.6561	.6571	.6580	.6590	.6599	.6609	.6618	1	2	3	4	5	6	7	8	9
4.6	.6628	.6637	.6646	.6656	.6665	.6675	.6684	.6693	.6702	.6712	1	2	3	4	5	6	7	7	8
4.7	.6721	.6730	.6739	.6749	.6758	.6767	.6776	.6785	.6794	.6803	1	2	3	4	5	5	6	7	8
4.8	.6812	.6821	.6830	.6839	.6848	.6857	.6866	.6875	.6884	.6893	1	2	3	4	4	5	6	7	8
4.9	.6902	.6911	.6920	.6928	.6937	.6946	.6955	.6964	.6972	.6981	1	2	3	4	4	5	6	7	8
5.0	.6990	.6998	.7007	.7016	.7024	.7033	.7042	.7050	.7059	.7067	1	2	3	3	4	5	6	7	8
5.1	.7076	.7084	.7093	.7101	.7110	.7118	.7126	.7135	.7143	.7152	1	2	3	3	4	5	6	7	8
5.2	.7160	.7168	.7177	.7185	.7193	.7202	.7210	.7218	.7226	.7235	1	2	2	3	4	5	6	7	7
5.3	.7243	.7251	.7259	.7267	.7275	.7284	.7292	.7300	.7308	.7316	1	2	2	3	4	5	6	6	7
5.4	.7324	.7332	.7340	.7348	.7356	.7364	.7372	.7380	.7388	.7396	1	2	2	3	4	5	6	6	7

$\log_{10}\pi = 0.4971,$ $\log_{10}2\pi = 0.7982$

Table of Common Logarithms (2)

	0	1	2	3	4	5	6	7	8	9	1	2	3	4	5	6	7	8	9
5.5	.7404	.7412	.7419	.7427	.7435	.7443	.7451	.7459	.7466	.7474	1	2	2	3	4	5	5	6	7
5.6	.7482	.7490	.7497	.7505	.7513	.7520	.7528	.7536	.7543	.7551	1	2	2	3	4	5	5	6	7
5.7	.7559	.7566	.7574	.7582	.7589	.7597	.7604	.7612	.7619	.7627	1	2	2	3	4	5	5	6	7
5.8	.7634	.7642	.7649	.7657	.7664	.7672	.7679	.7686	.7694	.7701	1	1	2	3	4	4	5	6	7
5.9	.7709	.7716	.7723	.7731	.7738	.7745	.7752	.7760	.7767	.7774	1	1	2	3	4	4	5	6	7
6.0	.7782	.7789	.7796	.7803	.7810	.7818	.7825	.7832	.7839	.7846	1	1	2	3	4	4	5	6	6
6.1	.7853	.7860	.7868	.7875	.7882	.7889	.7896	.7903	.7910	.7917	1	1	2	3	4	4	5	6	6
6.2	.7924	.7931	.7938	.7945	.7952	.7959	.7966	.7973	.7980	.7987	1	1	2	3	3	4	5	6	6
6.3	.7993	.8000	.8007	.8014	.8021	.8028	.8035	.8041	.8048	.8055	1	1	2	3	3	4	5	5	6
6.4	.8062	.8069	.8075	.8082	.8089	.8096	.8102	.8109	.8116	.8122	1	1	2	3	3	4	5	5	6
6.5	.8129	.8136	.8142	.8149	.8156	.8162	.8169	.8176	.8182	.8189	1	1	2	3	3	4	5	5	6
6.6	.8195	.8202	.8209	.8215	.8222	.8228	.8235	.8241	.8248	.8254	1	1	2	3	3	4	5	5	6
6.7	.8261	.8267	.8274	.8280	.8287	.8293	.8299	.8306	.8312	.8319	1	1	2	3	3	4	5	5	6
6.8	.8325	.8331	.8338	.8344	.8351	.8357	.8363	.8370	.8376	.8382	1	1	2	3	3	4	4	5	6
6.9	.8388	.8395	.8401	.8407	.8414	.8420	.8426	.8432	.8439	.8445	1	1	2	2	3	4	4	5	6
7.0	.8451	.8457	.8463	.8470	.8476	.8482	.8488	.8494	.8500	.8506	1	1	2	2	3	4	4	5	6
7.1	.8513	.8519	.8525	.8531	.8537	.8543	.8549	.8555	.8561	.8567	1	1	2	2	3	4	4	5	5
7.2	.8573	.8579	.8585	.8591	.8597	.8603	.8609	.8615	.8621	.8627	1	1	2	2	3	4	4	5	5
7.3	.8633	.8639	.8645	.8651	.8657	.8663	.8669	.8675	.8681	.8686	1	1	2	2	3	4	4	5	5
7.4	.8692	.8698	.8704	.8710	.8716	.8722	.8727	.8733	.8739	.8745	1	1	2	2	3	4	4	5	5
7.5	.8751	.8756	.8762	.8768	.8774	.8779	.8785	.8791	.8797	.8802	1	1	2	2	3	3	4	5	5
7.6	.8808	.8814	.8820	.8825	.8831	.8837	.8842	.8848	.8854	.8859	1	1	2	2	3	3	4	5	5
7.7	.8865	.8871	.8876	.8882	.8887	.8893	.8899	.8904	.8910	.8915	1	1	2	2	3	3	4	4	5
7.8	.8921	.8927	.8932	.8938	.8943	.8949	.8954	.8960	.8965	.8971	1	1	2	2	3	3	4	4	5
7.9	.8976	.8982	.8987	.8993	.8998	.9004	.9009	.9015	.9020	.9025	1	1	2	2	3	3	4	4	5
8.0	.9031	.9036	.9042	.9047	.9053	.9058	.9063	.9069	.9074	.9079	1	1	2	2	3	3	4	4	5
8.1	.9085	.9090	.9096	.9101	.9106	.9112	.9117	.9122	.9128	.9133	1	1	2	2	3	3	4	4	5
8.2	.9138	.9143	.9149	.9154	.9159	.9165	.9170	.9175	.9180	.9186	1	1	2	2	3	3	4	4	5
8.3	.9191	.9196	.9201	.9206	.9212	.9217	.9222	.9227	.9232	.9238	1	1	2	2	3	3	4	4	5
8.4	.9243	.9248	.9253	.9258	.9263	.9269	.9274	.9279	.9284	.9289	1	1	2	2	3	3	4	4	5
8.5	.9294	.9299	.9304	.9309	.9315	.9320	.9325	.9330	.9335	.9340	1	1	2	2	3	3	4	4	5
8.6	.9345	.9350	.9355	.9360	.9365	.9370	.9375	.9380	.9385	.9390	1	1	2	2	3	3	4	4	5
8.7	.9395	.9400	.9405	.9410	.9415	.9420	.9425	.9430	.9435	.9440	0	1	1	2	2	3	3	4	4
8.8	.9445	.9450	.9455	.9460	.9465	.9469	.9474	.9479	.9484	.9489	0	1	1	2	2	3	3	4	4
8.9	.9494	.9499	.9504	.9509	.9513	.9518	.9523	.9528	.9533	.9538	0	1	1	2	2	3	3	4	4
9.0	.9542	.9547	.9552	.9557	.9562	.9566	.9571	.9576	.9581	.9586	0	1	1	2	2	3	3	4	4
9.1	.9590	.9595	.9600	.9605	.9609	.9614	.9619	.9624	.9628	.9633	0	1	1	2	2	3	3	4	4
9.2	.9638	.9643	.9647	.9652	.9657	.9661	.9666	.9671	.9675	.9680	0	1	1	2	2	3	3	4	4
9.3	.9685	.9689	.9694	.9699	.9703	.9708	.9713	.9717	.9722	.9727	0	1	1	2	2	3	3	4	4
9.4	.9731	.9736	.9741	.9745	.9750	.9754	.9759	.9763	.9768	.9773	0	1	1	2	2	3	3	4	4
9.5	.9777	.9782	.9786	.9791	.9795	.9800	.9805	.9809	.9814	.9818	0	1	1	2	2	3	3	4	4
9.6	.9823	.9827	.9832	.9836	.9841	.9845	.9850	.9854	.9859	.9863	0	1	1	2	2	3	3	4	4
9.7	.9868	.9872	.9877	.9881	.9886	.9890	.9894	.9899	.9903	.9908	0	1	1	2	2	3	3	4	4
9.8	.9912	.9917	.9921	.9926	.9930	.9934	.9939	.9943	.9948	.9952	0	1	1	2	2	3	3	4	4
9.9	.9956	.9961	.9965	.9969	.9974	.9978	.9983	.9987	.9991	.9996	0	1	1	2	2	3	3	3	4

184

Table of Trigonometric Functions

	sin	cos	tan		sin	cos	tan
0	0.0000	1.0000	0.0000	45	0.7071	0.7071	1.0000
1	0.0175	0.9998	0.0175	46	0.7193	0.6947	1.0355
2	0.0349	0.9994	0.0349	47	0.7314	0.6820	1.0724
3	0.0523	0.9986	0.0524	48	0.7431	0.6691	1.1106
4	0.0698	0.9976	0.0699	49	0.7547	0.6561	1.1504
5	0.0872	0.9962	0.0875	50	0.7660	0.6428	1.1918
6	0.1045	0.9945	0.1051	51	0.7771	0.6293	1.2349
7	0.1219	0.9925	0.1228	52	0.7880	0.6157	1.2799
8	0.1392	0.9903	0.1405	53	0.7986	0.6018	1.3270
9	0.1564	0.9877	0.1584	54	0.8090	0.5878	1.3764
10	0.1736	0.9848	0.1763	55	0.8192	0.5736	1.4281
11	0.1908	0.9816	0.1944	56	0.8290	0.5592	1.4826
12	0.2079	0.9781	0.2126	57	0.8387	0.5446	1.5399
13	0.2250	0.9744	0.2309	58	0.8480	0.5299	1.6003
14	0.2419	0.9703	0.2493	59	0.8572	0.5150	1.6643
15	0.2588	0.9659	0.2679	60	0.8660	0.5000	1.7321
16	0.2756	0.9613	0.2867	61	0.8746	0.4848	1.8040
17	0.2924	0.9563	0.3057	62	0.8829	0.4695	1.8807
18	0.3090	0.9511	0.3249	63	0.8910	0.4540	1.9626
19	0.3256	0.9455	0.3443	64	0.8988	0.4384	2.0503
20	0.3420	0.9397	0.3640	65	0.9063	0.4226	2.1445
21	0.3584	0.9336	0.3839	66	0.9135	0.4067	2.2460
22	0.3746	0.9272	0.4040	67	0.9205	0.3907	2.3559
23	0.3907	0.9205	0.4245	68	0.9272	0.3746	2.4751
24	0.4067	0.9135	0.4452	69	0.9336	0.3584	2.6051
25	0.4226	0.9063	0.4663	70	0.9397	0.3420	2.7475
26	0.4384	0.8988	0.4877	71	0.9455	0.3256	2.9042
27	0.4540	0.8910	0.5095	72	0.9511	0.3090	3.0777
28	0.4695	0.8829	0.5317	73	0.9563	0.2924	3.2709
29	0.4848	0.8746	0.5543	74	0.9613	0.2756	3.4874
30	0.5000	0.8660	0.5774	75	0.9659	0.2588	3.7321
31	0.5150	0.8572	0.6009	76	0.9703	0.2419	4.0108
32	0.5299	0.8480	0.6249	77	0.9744	0.2250	4.3315
33	0.5446	0.8387	0.6494	78	0.9781	0.2079	4.7046
34	0.5592	0.8290	0.6745	79	0.9816	0.1908	5.1446
35	0.5736	0.8192	0.7002	80	0.9848	0.1736	5.6713
36	0.5878	0.8090	0.7265	81	0.9877	0.1564	6.3138
37	0.6018	0.7986	0.7536	82	0.9903	0.1392	7.1154
38	0.6157	0.7880	0.7813	83	0.9925	0.1219	8.1443
39	0.6293	0.7771	0.8098	84	0.9945	0.1045	9.5144
40	0.6428	0.7660	0.8391	85	0.9962	0.0872	11.4301
41	0.6561	0.7547	0.8693	86	0.9976	0.0698	14.3007
42	0.6691	0.7431	0.9004	87	0.9986	0.0523	19.0811
43	0.6820	0.7314	0.9325	88	0.9994	0.0349	28.6363
44	0.6947	0.7193	0.9657	89	0.9998	0.0175	57.2900
45	0.7071	0.7071	1.0000	90	1.0000	0.0000	∞

0795